Breeding Poultry For Egg Production

by Maine Agricultural Experiment Station

with an introduction by Jackson Chambers

This work contains material that was originally published in 1911.

This publication is within the Public Domain.

This edition is reprinted for educational purposes
and in accordance with all applicable Federal Laws.

Introduction Copyright 2018 by Jackson Chambers

The World's Largest Selection of Vintage Poultry Books

www.VintagePoultry.com

Self Reliance Books

Get more historic titles on animal and stock breeding, gardening and old fashioned skills by visiting us at:

http://selfreliancebooks.blogspot.com/

Introduction

I am pleased to present yet another title on Poultry.

The work is in the Public Domain and is re-printed here in accordance with Federal Laws.

As with all reprinted books of this age that are intended to perfectly reproduce the original edition, considerable pains and effort had to be undertaken to correct fading and sometimes outright damage to existing proofs of this title. At times, this task is quite monumental, requiring an almost total "rebuilding" of some pages from digital proofs of multiple copies. Despite this, imperfections still sometimes exist in the final proof and may detract from the visual appearance of the text.

I hope you enjoy reading this book as much as I enjoyed making it available to readers again.

Jackson Chambers

BULLETIN No. 192.

BREEDING POULTRY FOR EGG PRODUCTION.*

BY

RAYMOND PEARL.

PART I.

SUMMARY OF EARLIER WORK.

Since 1898 an investigation in breeding Barred Plymouth Rock fowls for increased egg production has been in progress at the Maine Station. This work was put under the direction of the present writer in 1907. No systematic or detailed report of the results obtained was made until 1909. Since that time a number of papers have been published dealing with one phase or another of these experiments. A list of these papers follows.**

1. Data on the Inheritance of Fecundity obtained from the Records of Egg Production of the Daughters of "200-egg" Hens. Maine Agl. Exp. Stat. Ann. Rpt. for 1909, pp. 49-84. (Bulletin 166).

2. A Biometrical Study of Egg Production in the Domestic Fowl. I. Variation in Annual Egg Production. U. S. Dept. Agr. Bur. Anim. Ind. Bulletin 110, Part I, pp. 1-80, 1909.

3. A Biometrical Study, etc. II. Seasonal Distribution of Egg Production. *Ibid.* Part II, pp. 81-170, 1911.

4. Is there a Cumulative Effect of Selection? Data from the Study of Fecundity in the Domestic Fowl. Zeitschr. f. ind. Abst. u. Vererbungsl. Bd. 2. pp. 257-275, 1909.

---

* Papers from the Biological Laboratory, Maine Agricultural Experiment Station, No. 32.

** Dr. Frank M. Surface, formerly Associate Biologist of this Station, is a joint author (with R. Pearl) of Nos. 1-5 inclusive of the papers listed. The other papers cited are by R. Pearl.

5. Studies on Hybrid Poultry. Maine Agr. Expt. Stat. Ann. Rpt. for 1910, pp. 84-116, 1910. (Pp. 100-106 deal with fecundity).

6. The Relation of the Results Obtained in Breeding Poultry for Increased Egg Production to the Problem of Selection. Rpt. 30th Meeting Soc. Proc. Agr. Sci. pp. (of reprint) 1-8.

7. Inheritance in "Blood Lines" in Breeding Animals for Performance, with Special Reference to the "200-Egg" Hen. Ann. Rpt. Amer. Breeders Assoc. Vol. VI, pp. 317-326, 1911.

8. Inheritance of Fecundity in the Domestic Fowl. Amer. Nat. Vol. XLV. pp. 321-345, 1911. (This paper is reprinted in full as Part IV of the present bulletin).

9. Biometric Arguments Regarding the Genotype Concept. Amer. Naturalist, Vol. XLV. pp. 561-566, 1911.

Most of these papers cited above deal with certain essentially negative results obtained in the earlier years of the experiments. It seems desirable, as an introduction to the positive results now reported, to review briefly the entire history of the work at the Maine Station in the experimental breeding of poultry with reference to the character fecundity or egg production. Another reason for publishing the present bulletin lies in the fact that most of the papers in which the original data and results have been presented are not easily accessible to the agricultural public. It is hoped that this bulletin may help to meet the demand for information on the part of that public in regard to the progress of the breeding work at the Maine Station. Finally it is desired to present at this time some data not hitherto published regarding obscure and doubtful points of interpretation and suggested criticisms of the experiments in breeding.

For detailed evidence on any point discussed, except such as are here presented for the first time, the reader is referred to the original papers listed above.

PLAN OF EARLIER WORK.

The earlier work of the Station on this subject, which covered the years 1898 to 1907 inclusive, was concerned, and executed in conformity, with the then prevailing views respecting the effectiveness of mass selection. The underlying idea which dominated these earlier experiments was that by breeding consistently year after year from the highest layers, regardless of all other considerations, there must be brought about a definite and steady, if gradual, improvement or increase in the average annual egg production per bird.

Two distinct and separate experiments were carried out during the period of the investigation prior to 1908. These may be designated as follows:

I. Experiment in continued selection of fluctuating variations in fecundity.

II. Experiment regarding the inheritance of fecundity.

I. *Experiment in continued selection of fluctuating variations in fecundity.* In 1898 there was begun at the Maine Agricultural Experiment Station an experiment to determine whether egg production in the domestic fowl could be increased by the continued selection of the highest egg producers as breeders. This experiment was planned and started by Director C. D. Woods and the late Professor G. M. Gowell.* An exact record was made of the egg production of each hen during the first year of her life; trap nests being used to furnish the individual records. The plan of the experiment begun in 1898 was to make from a strain of Barred Plymouth Rock hens, which had been "pure" bred, i. e., without introduction of strange "blood," for a long time by Professor Gowell, a continuous close selection with reference to egg production. The practice in breeding was to use as mothers of the stock bred in any year only hens which laid between November 1 of the year in which they were hatched and November 1 of the following year, 150 or more eggs. After the first year, all male birds used in the breeding were the sons of mothers whose production in their first laying year was 200 eggs or more. Since the normal average annual egg production of these birds may be taken to have been about 125 eggs, it will be seen that the selection practiced was fairly stringent.

Close inbreeding was not designedly practiced. It was always in theory possible to avoid this, since after the first four

---

* The present writer had *nothing whatever* to do with the planning of this experiment, nor with its conduct prior to December 1907. Therefore, he cannot justly be held accountable, as he has been by some critics, for real or supposed defects in the plan and earlier conduct of this experiment. The responsibility for the statistical analysis of the results is his, however.

years of the experiment the flocks were large (always containing more than 300 birds and usually nearer a thousand). While there was no close inbreeding no "new blood" was introduced into the strain from the outside during the period of the experiment.

II. *Experiment regarding the inheritance of fecundity.* In 1907 the experiment described above, having led to definite results was brought to an end. There was planned for 1908 a new experiment designed to test from another standpoint the conclusions which had been tentatively reached from the earlier experiment. In the conducting of the long selection experiment the females used as breeders were grouped into two classes, viz., (a) "unregistered" or birds laying 150 to 199 eggs in the pullet year, and (b) "registered" or birds laying 200 or more eggs in the pullet year.

It had been noted that the daughters of the so-called "registered" hens (namely hens that had produced 200 or more eggs each in the pullet year) did not usually make high egg records. The "200-egg" birds which made up the "registered" flock came, in most instances, from the "unregistered" mothers.

Experiment II was planned primarily to answer the following question: Will the daughters of high laying hens ("200-egg" birds) on the average produce more eggs in a given time unit than will birds of less closely selected ancestry?

The experiment was carried out according to the following plan: On the first of November, 1907, there were put into house No. 2, of the Station plant, 250 pullets. Each of these was the daughter of a hen that had laid approximately 200 eggs in her pullet year. As a matter of fact 11 of the 33 hens which produced these 250 "registered" pullets had each laid a few eggs less than 200 in a year forward from Nov. 1 of their pullet year. The writer has been criticized for including these birds in the work. When carefully considered such criticism appears to be without any real significance. In the first place nearly all of these 11 birds *were* "200-egg" hens in the sense that they laid this number of eggs (or more) in a period of 365 days following the laying of their first egg. The records were for the sake of uniformity in presentation and analytical discussion in (1)* and (4) taken as from November 1 of the pullet year to November 1 of the next year. That the records are taken in this way in no wise interferes with the fact that these birds were heavy layers. The further fact which entirely suffices to justify the inclusion of these 11 birds with the 22 which laid 200 or more eggs in the year from November 1, flows from the comparison of the *daughters* of the 11 with the *daughters* of the 22 in respect to average egg production. Table I of (1) shows that the mean winter production of all "registered" pullets was 15.29. The mean winter production of the 67 daughters of the 11 mothers under discussion was 16.03, and the mean winter production of the 125 daughters of the other 22 mothers was 14.87. So far, then, from the low average winter production of all "registered" pullets in this experiment taken together being due to the inclusion of these 11 mothers, whose November 1 to November 1 record fell a little below 200 eggs, and their daughters, actually this group of progeny had a *higher* winter production than the remainder of the "registered" flock.

These pullets were divided into flocks of 50 each and were fed and handled in every way exactly alike. At the same time that these 250 "registered" pullets (so-called because from "registered" mothers), were put into the house there were also put in 600 other Barred Plymouth Rock pullets. These other pullets were of approximately the same age as the 250 "registered" pullets and differed in their breeding only in respect to their mothers. They came from hens that had laid less than 200 eggs during the pullet year and more than 160. "Registered" cockerels (from the "200-egg" line) were used as the male parents for all the pullets both "registered" and "unregistered." The 600 "unregistered" birds were divided into flocks as follows: Two flocks of 50 birds each were kept in two pens in house No. 2, exactly like the pens in which the "registered" birds were kept. The remaining 500 birds were divided into four flocks—two of 100 birds each and two of 150 birds each and housed in the four pens of house No. 3. These pens are essentially like those of house No. 2, differing chiefly in the

---

* Figures in parenthesis refer to the papers in the list of literature at the beginning, p. 113.

matter of size. A trap nest* record was kept of the exact individual egg production of each of these birds.

RESULTS OF EARLIER WORK.

The essential results of the two lines of investigation described above may be very briefly set forth here. Greater details are given in papers (1-8) cited above. These results are:

1. That mass selection for high egg production on the basis of the trap nest record of the individual alone *did not*, as a matter of fact, result in a steady, continuous improvement in average flock production, though it was continued for a period of ten years.

2. That, as a matter of fact, the daughters of "200-egg" hens with from 6 to 9 years of mass-selected ancestry (on the basis of trap nest records) behind them were *no better layers* on the average than birds bred from the general flock.

Now whatever opinion anyone may hold as to the biological interpretation of these results he must not after all forget that they are *facts*. While it has been argued that 10 years is far too short a time to learn anything about the effect of selection it should be remembered that he who makes this argument is really discussing a very complex *theoretical* matter. An unbiassed examination of the literature on the subject indicates that the length of time which is considered necessary to prove experimentally the effectiveness or non-effectiveness of mass-selection depends almost entirely upon which way the results are coming. If after following a plan of mass selection for even 3 or 4 years one finds that concurrently there has been an improvement in the character selected for, he almost invariably and quite humanly concludes that the selection is the *cause* of the improvement. Just why, however, *post hoc* should be considered to be *propter hoc* when it happens to be "your" *hoc* but not at all so when it is "my" *hoc* that is concerned has never been clear to the writer. It certainly seems fair to suppose that it requires just exactly as many years *critically* to prove by experiment that mass selection in a particular case *is* effective

---

*For a description of the trap nest used in the breeding work of the Station see "Appliances and Methods for Pedigree Poultry Breeding" by R. Pearl and F. M. Surface. Me. Agric. Exp. Station. Bulletin 150. pages 239-274. 1908.

as it does to prove in another case that it is *not* effective. The situation here is precisely as broad as it is long.

Practically, from the standpoint of the plain poultryman, whose interest in poultry keeping is confined to some part of the span of an ordinary lifetime, these results at the Maine Station give little encouragement to the idea of wholesale trapnesting with the expectation of thereby increasing the egg production of the flock. That the trap nest has a place in poultry husbandry is certain. It is equally certain, however, that trapnesting for the purpose of improving egg production by the selection of the best layers has not that degree of practical usefulness and importance which it was popularly supposed to have some ten years ago when the work of the Maine Station in breeding for egg production was being so extensively exploited by the agricultural press and by institute workers all over the country. It seems now to be quite generally agreed that about the only profitable function of the trap nest in practical or commercial (as distinguished from experimental) poultry keeping is in connection with special needs or problems, as for example, in the work of the fancier, who desires to keep individual pedigrees of his stock. There does not exist any critical evidence that the selection of the highest laying birds on the basis of the trap nest record as breeders will insure or guarantee any definite, permanent improvement in average flock production.

Since as a matter of fact, as the work at this Station shows, this method of selecting breeders has very little, if any, real relation to the average production of subsequent flocks, it is obvious that, as a mere matter of chance, temporary improvement in production might be expected to follow this plan of breeding in about 50 percent of all flocks on which it was tried, and a temporary decline in production in the other 50 percent. This appears to be the actual state of the case. Some practical poultrymen who have tried trapnest selection of the best layers as breeders have obtained improved average egg yields for a time at least. They attribute the improvement to the selection, though without any critical evidence, of course, and are enthusiastic believers in the gospel of the trap nest. Other equally competent poultrymen have failed to get any such improvement and have discarded the trap nests, though sometimes, it must be confessed, clinging firmly to the theory of breeding which their own experience has shown to be at least *practically* inadequate to meet their needs.

Not only was there no improvement in average flock production following the method set forth in the preceding section, but actually there was a slight decline in production during the selection period. No particular importance, however, is, in the writer's opinion, to be attached to this decline. It probably is to be regarded as due to chance, i. e., to a number of accidental causes operating together. (See p. 156 for further discussion of this matter.)

The results of this earlier work aroused a good deal of protest and criticism on the part of ardent believers in the efficacy of mass selection under all circumstances. Furthermore many persons have offered tentative explanations as to why these experiments in selecting for improved egg production resulted as they did. Some of these suggested criticisms and explanations have been published, but most of them have not, but instead have been confined to verbal discussions among workers interested in the problems of breeding.

No attempt has been made by the writer to answer criticisms of this work.* The discussion which follows has no polemic object. When, as in the present case, the point at issue is the critical interpretation of admitted results which are (and must be in nearly all cases) in some degree incomplete no amount of argumentation as to what "might" or "ought" to obtain, really helps very much in getting at the true facts. The most useful course would seem to be first to examine critically all possible interpretations and then devise if possible ways of testing experimentally which, if any, of these interpretations are really valid. With the presentation of the evidence so obtained the scientific case must rest, it seems to me, until additional and directly pertinent evidence can be brought forward. While the search for data bearing critically on the interpretation of breeding experiments on fecundity at this Station is by no means completed, yet it seems desirable now that certain of the positive results of the later experiments are to be presented to consider critically the possible interpretations of the earlier work, and to bring forward some of the evidence which has led the writer to the opinion which he holds.

---

*With the exception of the paper numbered 9 on the list at the beginning. That paper deals only with a few special points.

PART II.

Critical Consideration of Possible Interpretations of Earlier Work.

The critical interpretation of the results of the mass selection experiment described in the preceding section is by no means a simple matter. As to the bare facts as such there can be no question, but how shall they be interpreted? What really do they mean?

There are two principal general interpretations or explanations which may conceivably be given for the selection experiments at the Maine Station between 1898 and 1907 turning out in the way which they did. On the one hand it may be said that the results indicate that the general theory of the effectiveness of selection, or even more broadly the theory of breeding, which was at the foundation of this experiment, is, in greater or less degree, inadequate or incorrect. That is to say, the experiment may be interpreted, as it has been by the writer, as showing that it is doubtful whether the picking out by selection of minute favorable variations has in reality any cumulative or additive effect, so far as concerns the hereditary or germinal constitution of an animal, at least with reference to the character fecundity or egg production in fowls.

Before reaching such a conclusion, however, one must consider on the other hand, alternative interpretations and see whether the facts cannot be equally well or better explained in some other way. A number of such alternative explanations may be thought of. Nearly all of these explanations which suggest themselves fall into one category. This category is, to characterize it in a word, *the effect of environment*. In general terms this explanation of the results obtained would run something like this: that in reality the selection for increased egg production practiced during the years 1898-1907 was inherently or potentially effective, but that during this same period of years one or another or a combination of environmental circumstances became progressively worse, so that the gain which may be supposed to have been made each year as a result of the selection was masked or hidden by the untoward effect of the environment which prevented the hens from laying up to what was their true or innate capacity in the way of fecundity.

Specifically the possibilities here are large. There are many sorts of things by which a hen's laying may be disturbed and reduced. The action of such environmental circumstances furthermore cannot be prevented nor their disturbing influence upon a selection experiment eliminated by "keeping the environment constant" during the course of the experiment. This, of course, is the usual experimental method of attempting to safeguard against environmental factors disturbing the interpretation of the results of an experiment having to do with inheritance. But, as a matter of fact, leaving aside as of no real importance in the present discussion the fact that with such animals as poultry certainly it is physically impossible to obtain anything more than *average* uniformity of environment during a long period of years, there is a further point not to be lost sight of. This is that the effect of any *adverse* environmental circumstance acting upon an animal during the course of a long continued experiment in selection *must tend to become progressively cumulative as time goes on, if it be really efficiently adverse at all.*

What is meant is this: Suppose at the outstart of the experiment something in the method of feeding, or in the method of incubation, or of rearing the chicks was of a character such as to affect adversely, even to a slight degree, the vitality or constitution of the birds. Even without any true inheritance of this effect nevertheless its action must necessarily tend to become cumulative for purely physiological reasons, because (to confine the discussion to the case in hand, namely the domestic fowl) it admits of no question that a constitutionally weak or debilitated fowl lays an egg which is "weak" also. The elaboration of the yolk and of the albumen takes place within the hen's body. These substances serve as the food of the developing embryo. It is certain from observation of both egg and chick that the same kind or quality of food is not furnished to the embryo by the egg manufactured in the body of a strong fowl as is furnished in an egg manufactured in the body of a weak fowl. This is a fact which is well known to everyone who has had experience in the hatching and rearing of poultry. To analyze minutely all of the biological and chemical factors involved would certainly be a very difficult, indeed an almost impossible task, yet because such analysis is not easily possible in no wise militates against the fact itself.

Furnished with a qualitatively inadequate food supply the developing embryo either dies before hatching or hatches into a weak, debilitated chick. This badly nourished, weak chick grows into an adult fowl which is weak in constitution; usually weaker and to a greater degree lacking in vitality than the parent. The reason is that the unfavorable environmental factor has had a double action upon the adult offspring. Not only did it start life as an improperly nourished weak embryo, but throughout its post-embryonic development to the adult condition the same unfavorable environment which acted adversely upon its mother has been acting upon it and undoubtedly with increased efficiency because of the initial weakness of the embryo. This offspring bird may thus be expected to produce a still less normal supply of nutriment in its eggs than did *its* mother, since it is less vigorous and normal than she was.*

Thus the weakness is passed on from generation to generation tending all the time to become greater. I think that it must be obvious in view of these considerations that any environmental condition which is adverse to general constitutional vitality, if it is effective at all, must tend to become cumulatively so, even though every effort be made to keep environmental conditions uniform during the experiment. In fact the more uniform the environment is kept the more certainly will there be a cumulative effect of any unfavorable factor in it.

Obviously such a result as that under discussion has no real relation to the problem of the inheritance of acquired characters, though the objective result itself is precisely that which would be expected if a weakness induced by the environment were inherited. But actually the factor here dealt with is a purely nutritional one, and has nothing whatever to do with germinal constitution. This fact that any adverse environmental factor tends to produce an effect on the organism (at least among birds and mammals) which is persistent and in greater or less degree progressively cumulative, so long as the environment is

---

*Certain of these matters are being made the object of special investigation from a practical standpoint by Prof. Rice of Cornell University. Cf. Rice, J. E., and Rogers, C. A., Importance of Constitutional Vigor in the Breeding of Poultry. Cornell Reading Course for Farmers. No. 45. 1909 pp. 777-796.

kept constant and the factor continues to act, is, of course, one reason why it is so exceedingly difficult to get really critical evidence on the question of the inheritance of acquired characters.

In addition to the cumulatively adverse effect of environment as an explanation of the results of the earlier work at this Station another possible interpretation which is essentially physiological in its nature occurs to one. It is that any favorable progress in the way of increasing egg production by the selection was offset in the experiment by the weakening and debilitating effect upon the birds of the inbreeding which it might be contended was practiced during the experiment (see p. 115 above for facts on this point).

Another suggestion which has been made is that while there was no progressive increase in egg production following the mass selection this has no bearing on the question of the effectiveness of the selection of minute fluctuating variations because the character fecundity or egg productiveness is not inherited in the domestic fowl at all.

It is the purpose of this section to discuss these various suggestions one by one, presenting evidence which it is hoped will help to throw light on the subject. The evidence on the last mentioned criticism (that having to do with the non-inheritance of fecundity) will be presented in Part IV.

ARTIFICIAL HATCHING AND REARING.

The practical management of poultry naturally and obviously divides itself into three great divisions, namely (1) housing, (2) feeding, (3) incubating and rearing the chicks to replenish the flock. In considering the possibility that in the experiments under discussion some adverse factor in the environment (which broadly-speaking in the case of poultry on an intensive plant means the management) masked or concealed a favorable effect of the selection, it will be well to deal with each of these three divisions of management separately.

The methods of *housing* or *feeding* the stock practiced during the period of the selection experiment cannot, I think, reasonably be held to have had any adverse effect upon, or to have masked or covered up, any innate, inherited improvement in egg production conceived to have resulted from the selection. The reasons why this would seem to be the case are as follows: The system of housing in the so-called "curtain front" type of house which has been used at the Maine Station practically from the beginning of these experiments has been widely adopted by practical poultrymen all over this country and indeed in all parts of the world. It was the first attempt at the "fresh air" principle in housing poultry and has, with some modifications in recent years, grown steadily in favor in the minds of practical poultrymen who know that this method of housing, so far from adversely affecting egg production instead actually promotes a better egg production because it helps to keep the birds in a better general physiological condition than in any type of house yet devised.*

The same consideration applies with reference to the method of feeding used throughout the selection experiment. The Maine Experiment Station's dry mash has been very widely used indeed as a laying feed. It is a feed calculated for, and in actual practice shown to be excellently adapted to, stimulating the birds to something approaching the maximum of production of which they individually are constitutionally or hereditarily capable.*

This brings us to a consideration of the third of the great divisions of poultry management, namely, the hatching and rearing of the chickens. Here there are two methods: (a) the natural, in which the eggs are incubated by a brooding hen and the chickens are reared by a hen; and (b) the artificial, in which the eggs are incubated in incubators and the chickens are reared in brooders. Each of these methods is widely practiced and each has its staunch adherents. There is, however, a very wide spread feeling amongst poultrymen that artificial hatching and rearing has, in the long run, an injurious effect upon the stock. This injurious effect, they will grant, may not be apparent at once. But if artificial methods are persisted in for a long period of time those holding this view maintain that the

---

*This statement is intended to include "fresh air" houses of the Tolman and other patterns in the same *general* category as the "curtain front" house.

*This, of course, does not mean that *any* of the records of egg production ever obtained at the Maine Station approach the *physiological* limits of fecundity of the domestic fowl as a *class*. Cf. on this point (9).

result will be a steady and definite, if gradual, deterioration of the stock in respect to vitality or constitutional vigor and productiveness.

Now since a time previous to the beginning of the selection experiment at this Station in 1898 no chicken has ever been hatched on the Experiment Station plant except in an incubator nor reared in any other way than in a brooder. That is to say, the flock of hens on the Experiment Station plant in 1911 represents the end link in an unbroken chain of more than 13 years (which here mean "generations") of continuous artificial incubation and artificial rearing. It is quite evident, I think, that if these processes do bring about deterioration of the stock in vitality and productive qualities such deterioration ought by this time to be beginning at least to make itself apparent.

The possibility that such a deterioration in vitality and productiveness due to continued artificial incubation and rearing was the real reason why during the period of mass selection from 1898 to 1907 there was no improvement in the average egg production of the flock but instead a slight decrease certainly demands careful consideration. This interpretation of the results has specially appealed to a number of poultrymen. Thus Dryden * says (*loc. cit.* p. 382): "In the nine years breeding work at the Maine Station artificial methods were used in hatching and brooding the chicks, and while we are guessing at the failure to secure high egg yield in this experiment I venture to guess the failure was due to a gradual lowering of vitality in the stock by artificial incubation."

The question then to be considered, in light of all the available facts, is as to whether there was during the course of the experiment in selection any lowering of vitality due to this cause, and further whether this can be regarded as the explanation of the failure of an increase in average egg production to appear during the selection period.

From the nature of the experimental work which it was desired to do it was impossible practically to employ natural methods of hatching and rearing when the writer took charge

---

* Dryden, J. Artificial vs. Natural Incubation. Rpt. Am. Breeders' Association. Vol. V. pp. 380-382, 1909.

of the work in 1907. Artificial incubation had to be continued. It would seem that there are two lines of approach to the question as to whether the failure of the selection experiment to result in increased yields was due to deterioration following artificial incubation. On the one hand may be considered the actual facts regarding other evidence of deterioration besides egg production. While egg production is one index of vitality and constitutional vigor it is by no means the only one. Mortality and morbidity are other indices; so also is the hatching quality of eggs. Did the flock show evidence of a real constitutional degeneration, as indicated not alone by egg production, but by these other factors as well?

In considering this whole question it should be recognized that there is a difference between real constitutional degeneration and merely a state of temporary low condition due to an unfavorable immediate environment. The one is permanent and the other is only transitory. The one is truly constitutional, the other superficial. This brings us to the consideration of the second line of evidence which it is possible to get on the problem under discussion. If the real cause of the persistently low egg production during the selection experiment was artificial hatching and rearing merely changing the method of breeding without changing the method of incubation or rearing would certainly be expected to produce no effect. If a purely *environmental* matter such as artificial hatching and breeding is an efficient check to improvement by one method of breeding, it ought if unchanged to act with equal effectiveness against any other system of breeding. If it does not so act one is forced to the conclusion that it was not really an effective factor in determining the results of the first method.

Let us now turn to the data. In regard to the first line of evidence, namely facts presented by other indices of general vigor and vitality besides egg production, adult mortality may be first considered. Table A gives the number of adult females put in the laying house, the number of these which died and the percentage mortality for each of the years covered by the mass selection experiment. The "number of adult females" means the number of pullets put into the laying house each year at the average age of about 6 to 7 months. They are the same birds whose egg records during the same period are shown in fig. 80, (p. 157).

TABLE A.

Mortality Records of Adult Females During the Period 1898 to 1907.

Laying year.	Total number of birds put in house.	Number which died during year.	Percentage mortality.
1899–1900	81	8	9.9
1900–'01	100	14	14.0
1901–'02	55	3	5.5
1902–'03	160	13	8.1
1903–'04	300	44	14.7
1904–'05	550	33	6.0
1905–'06	700	64	9.1
1906–'07	700	44	6.3
1907–'08	850	69	8.1

The percentage mortality data are shown graphically in fig. 76. The straight line is the graph of the equation $y = 11.18 - 0.42x$ where y denotes percentage mortality and x number of years since 1898–'99. The line is fitted to the observations by the method of least squares.

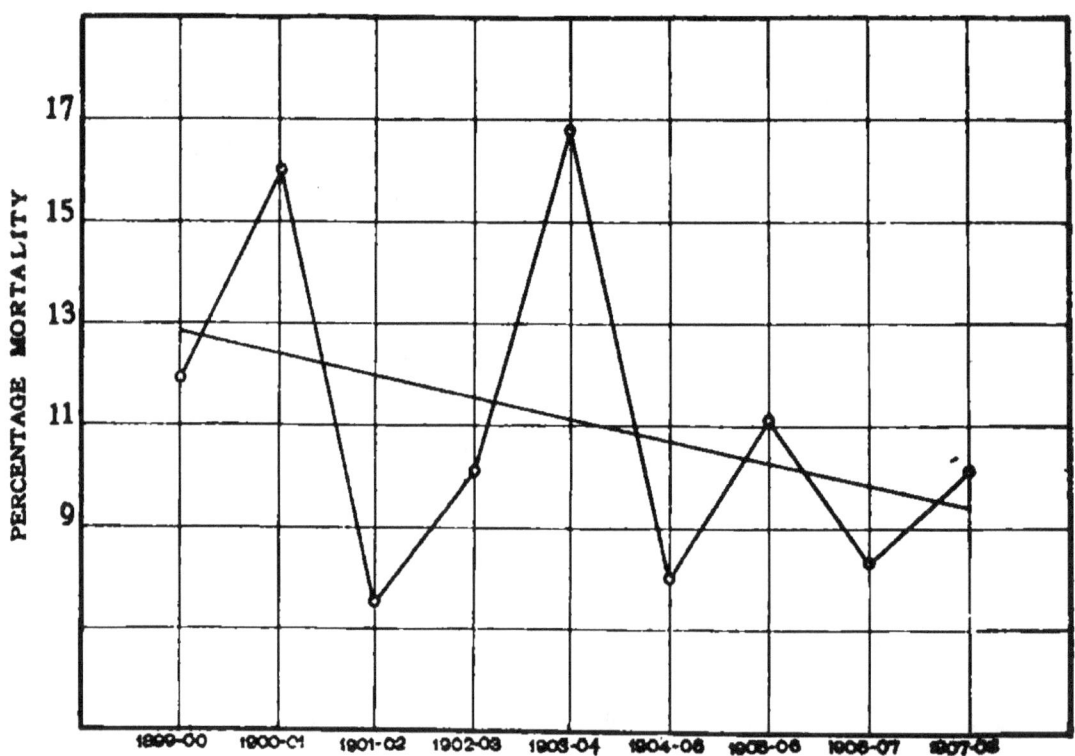

FIG. 76. Diagram showing the percentage mortality of adult female birds during the course of the selection experiment.

From the table and diagram the following points are to be noted:

1. As would be expected, the percentage mortality is seen to fluctuate in amount from year to year. These variations are without doubt to be accounted for by differences in general environmental factors in different years, and to accidents.*

2. The general trend, however, of the percentage mortality is plainly *downward* during the period covered by the experiment in mass selection. That is, during the period from 1899 to 1908, when the egg production was showing a slightly downward trend the adult mortality was also distinctly diminishing.

3. There is no evidence from the figures of adult mortality to indicate either that artificial incubation and rearing or any other environmental condition adversely affected the constitutional vigor of the strain during the course of the mass selection experiment, or (b) that in such action is to be found in the explanation of the failure of that experiment to result in increased annual average egg production.

The percentage mortality figures are given here only for the period covered by the selection experiment in order that direct comparison may be made between the trend of the egg production during that period (see fig. 80, p. 157) and the mortality curve. It may be said, however, that since the laying year 1907-'08 the mortality of adult birds has fluctuated about the same average as during the three years preceding. During the laying year 1910-11 the adult mortality has been quite exceptionally low.

Having now considered adult mortality as an index of general vitality and constitutional vigor attention may next be turned to the records of the hatching quality of eggs and the mortality of chicks. It needs but little practical acquaintance with poultry for anyone to recognize that in these two things exists one of the most precise measures of the general vitality or constitutional vigor of a strain or flock that it is possible to get. If a relatively large percentage of eggs hatch, and the chickens are strong and vigorous, and only a small proportion of them die it puts the question of the vitality of the stock beyond cavil.

---

* Cf. (2), pp. 16-19.

It would be highly desirable if records could be presented for the hatching of eggs and the mortality of chicks during the whole period covered by the mass selection experiment. Unfortunately it is impossible to do this since no such records were kept prior to the hatching season of 1908 (laying year 1907-08). Data on these points have been kept since that time, however. The figures for the hatching seasons of 1908 and 1909 have already been published.* For comparison with these, and to serve as a basis for a discussion of the present condition of the stock, in respect to these matters and in relation to the method of breeding now being followed, similar data will be given here for the hatching season of 1911 (laying year 1910-11).

Tables B and C give the hatching records for the season of 1911 for birds belonging respectively to high and low fecundity lines. All of the birds in Table B belong to pedigree lines in which the mean winter egg production of all females for four generations has been high. These are the birds used in 1911 to propagate the "high lines," the egg production of which is discussed farther on. All of the birds in Table C belong to pedigree lines in which the mean winter production has been uniformly low for four generations. The females in these two tables represent of course only a part of all birds bred in 1911, but they are the only ones whose hatching records are pertinent to the present discussion. Cross-bred birds and Barred Rocks in other lines of experimentation do not concern us here. As a matter of fact, the averages of Table B represent about the average hatching and rearing record of all birds bred in 1911. Some groups of birds did conspicuously better, especially on the rearing records.

---

* Pearl, R. and Surface, F. M. Studies on the Physiology of Reproduction in the Domestic Fowl. IV. Data on certain Factors Influencing the Fertility and Hatching of Eggs. Ann. Rpt. Me. Agr. Expt. Stat. for 1909, pp. 105-164.

From these tables the following matters are to be noted:
1. The percentage of infertile eggs is distinctly high, being

TABLE B.

Hatching Records of Barred Rock Females of High Fecundity Lines. Season of 1911.

Hen No.	Eggs set.	Infertile.	Per cent infertile.	Died in shell.	Hatched.	Per cent fertile eggs hatched.	Chicks died in 3 weeks.	Per cent died in 3 weeks.
4	43	2	4.6	15	26	63.4	3	11.5
9	57	10	17.5	19	28	59.5	3	10.7
10	69	20	28.9	20	26	53.1	1	3.8
11	18	5	27.7	6	7	53.8	0	0
12	60	2	3.03	10	48	82.7	6	12.5
14	64	14	21.8	8	42	84.0	3	7.1
18	43	7	16.2	34	2	5.1	0	0
19	48	3	6.2	13	32	71.1	2	6.2
20	45	1	2.2	24	20	45.4	3	15.0
22	38	4	10.5	32	2	5.8	1	50.0
25	53	7	13.2	22	22	47.8	2	9.1
27	57	19	33.3	25	13	34.2	0	0
28	45	4	8.8	15	25	60.9	2	8.0
29	53	9	16.9	14	29	65.9	3	10.3
36	47	4	8.5	11	32	74.4	4	12.5
39	43	8	18.6	33	2	5.7	0	0
41	51	0	0	30	20	39.2	0	0
46	40	30	75.0	5	4	40.0	1	25.0
53	38	8	21.1	18	12	40.0	2	16.6
62	52	21	40.3	13	15	48.3	3	20.0
76	46	7	15.2	8	31	79.4	4	12.9
77	34	33	97.1	1	0	0	0	0
81	37	3	8.1	6	27	79.4	1	3.7
85	56	15	26.7	27	13	31.7	2	15.3
117	50	4	8.0	13	33	71.7	5	15.1
134	50	28	56.0	18	4	18.1	0	0
165	38	3	7.8	11	24	68.5	1	4.1
196	42	10	23.8	12	20	62.5	5	25.0
198	46	5	10.8	24	17	41.4	3	17.6
273	46	22	47.8	7	16	66.6	1	6.2
1138	35	2	5.6	3	29	87.8	2	6.8
1156	37	13	35.1	8	10	41.6	4	40.0
1169	42	6	14.2	24	11	30.5	2	18.1
1170	40	0	0	5	35	87.5	2	5.7
1191	51	2	3.9	25	23	46.9	13	56.5
1196	42	5	11.9	6	30	81.1	0	0
1203	40	5	12.5	16	19	54.2	4	21.0
1213	48	3	6.2	22	24	53.3	10	41.6
1225	40	6	15.0	25	9	26.4	2	22.2
1232	39	1	2.5	13	24	63.1	4	16.6
1255	21	10	47.6	7	4	36.3	0	0
1256	42	3	7.1	11	28	71.7	2	7.1
Totals and weighted averages	1886	364	19.3	659	838	55.1	106	12.6

19.3 percent in one case and 18.8 percent in the other. This, however, does not mean, as might at first thought be supposed that there is some weakness on the part of the males or females bred. As a matter of fact the explanation of this poor record of fertility is that when the pens were mated up early in February no interval of time was allowed for the establishment of fertility before beginning to save eggs for hatching. That is to say, the eggs were saved and incubated from these breeding pens on the very same day that the male bird was put in the pen, and in a few instances it is probable that eggs were actually included from hens which had never been with a male bird at all. Every practical poultryman knows that on the average it takes from 6 to 10 days to establish good fertility in eggs

TABLE C.

Hatching Records Barred Rocks Females of Low Fecundity Lines. Season of 1911.

Hen No.	Eggs set.	Infertile.	Per cent infertile.	Died in shell.	Hatched.	Per cent fertile eggs hatched.	Chicks died in 3 weeks.	Per cent died in 3 weeks.
3	27	1	3.7	15	11	42.3	1	9.1
34	39	2	5.1	17	20	54.0	2	10.0
45	49	22	44.8	12	15	55.5	3	20.0
55	53	36	67.9	13	4	23.5	1	25.0
56	55	4	7.2	32	19	37.2	1	5.2
118	43	0	0	21	21	48.8	4	19.0
142	38	2	5.2	23	13	36.1	1	7.6
188	69	2	2.8	23	43	64.1	0	0
221	57	12	21.1	36	9	20.0	3	33.3
232	68	9	13.2	32	27	45.7	3	11.1
249	32	5	15.6	14	13	48.1	3	23.0
430	53	34	64.1	2	17	89.4	3	17.6
477	38	1	2.6	25	12	32.4	1	8.3
479	54	11	20.3	18	25	58.1	8	32.0
481	59	2	3.3	24	33	57.8	3	9.1
546	48	15	31.2	27	6	18.1	1	16.6
1221	43	3	6.9	22	17	42.5	5	29.4
1238	49	4	8.1	22	23	51.1	7	30.4
Totals and weighted averages	874	165	18.8	378	328	46.2	50	15.2

after birds are mated together. As a result of this incubating of the eggs taken from the time that the male and female birds were put together the record of fertility suffers a heavy handicap. Actually after fertility was once established (that is, after the male bird had been in the pen about 10 days) the average percentage of infertile eggs for the remainder of the season was about that which is considered normal in the work of this Station. This figure is on the average about 10 percent of the eggs infertile.

The reader may be disposed to wonder why eggs which were practically certain to be infertile were incubated. The reason was primarily that it was desired to get just as many chicks as possible hatched April 1 or within a few days of that time. Experience has shown that, under the environmental conditions which obtain here, that time is the best to hatch chickens which are to be used in fecundity work. Such birds come into laying at the proper time without either forcing or retarding. Now it is a fact that while, on the average, it takes from 6 to 10 days to get fertility well established after a mating is made, yet there are individuals in which the very next egg laid after the first copulation will be fertile. Because of this consideration all possible eggs were saved from the beginning of the matings, with the certain knowledge that while the *relative* or percentage fertility of these early eggs would be low, yet *absolutely* a few chicks would be obtained. The desire to get these chicks far outweighed any idea of making a maximum record of fertility of eggs, the latter, in fact, not entering into consideration at all.

2. The hatching quality of the eggs as indicated by the percent of fertile eggs hatched is again somewhat below what may be considered normal for the Maine Station stock at the present time. With large numbers of eggs the normal hatching percentage of fertile eggs is on the average a little over 60, taking the whole season through. Toward the last of the mating season (the month of May) the hatching percentage normally rises considerably.

Here, just as in the case of fertility, the records tabled bear a rather heavy handicap, which could have been avoided had the only purpose been to bring out the best record of which the birds were capable. The factor in question here is the holding of the eggs before incubation from the first week in February on. No eggs were put in incubators until March 7. More than half of the eggs set at this time were over two weeks old when put in the incubator. Everyone who has dealt practically with incubation knows that this means a serious reduction in the percentage of fertile eggs hatched. The reason for managing in this way was again to get the greatest possible *absolute* number of chicks hatched about April 1, regardless of the *relative* proportion of chicks to eggs set.

3. Taking all the records together and using the averages in the computations it appears that, even with the handicaps mentioned, *in the high fecundity lines it required only 2.6 eggs in the incubator in 1911 to produce one chicken three weeks old. In the low fecundity lines it required 3.2 eggs to make one*

chicken three weeks old.* These figures are for the whole of the hatching season of 1911, that is, from February 1 to June 1. They do not represent the normal reproducing ability of the stock because of the heavy handicap explained above. In spite of this fact, however, these records can only be regarded as indicating an excellent performance.

Certainly these figures for hatching and rearing give no support to the view that the constitutional vigor or vitality of the Station Barred Plymouth Rock stock has been impaired by many years of continued artificial incubation and rearing. When it takes but three eggs or less to produce a chick three weeks old the stock cannot be said to be in a condition of reduced vitality.

4. It is plain that there is no substantial difference between the females of the high fecundity lines and the females of the low fecundity lines with respect to hatching records. What small differences there are indicate that birds of the high fecundity lines are on the whole somewhat surer reproducers than those of the low fecundity lines. While the percentage of infertile eggs is smaller in the low fecundity lines, on the other hand the percentage of fertile eggs hatched is also lower and a slightly larger percentage of chicks died during the first three weeks of their lives. Particular attention is called to this matter here because it has been alleged by one critic that selection for high egg production was inimical to reproductive capacity in the domestic fowl. As a matter of fact, as the present figures show, this is not at all the case. The criticism was based upon a fact previously brought out ** that there is a negative correlation between winter egg production and the hatching quality of eggs in the subsequent breeding season. This, however, is purely a physiological and not a genetic matter. High laying during the winter months undoubtedly tends to bring about a somewhat fatigued condition of the whole reproductive system with the result that the eggs in the subsequent spring do not hatch quite so well as under other circumstances. This, however, has nothing to do with the innate hereditary capacity of these same birds in respect to fecundity. This fact is indeed so

---

* As is well known three weeks covers nearly the entire chick mortality. The subsequent death rate among chicks which at three weeks of age are in full health and vigor is relatively insignificant.

** Cf., Pearl and Surface, *loc. cit.*

evident from Tables B and C as not to require further discussion.

We come now to the consideration of the last point in connection with artificial incubation. As was pointed out above (p. 127), if this had been the cause of the failure to increase egg production during the mass selection experiment, it ought to act as an equally efficient cause to prevent increase of egg production by any other method of breeding. As a matter of fact it did not so act, as is proven by the data given in Part IV of this bulletin. Without any change whatever from artificial methods of incubation or rearing it has been possible to isolate and breed from the Maine Station stock strains or lines in which *high* fecundity has been maintained during four generations at least. If the reader will study the diagram on p. 168 (fig. 84) remembering that during the *whole* period covered by the diagram there has been nothing but artificial incubation and rearing practiced he will find it difficult to believe that this factor has influenced, either one way or another, the results of the breeding experiments on fecundity at the Maine Station.

INBREEDING.

It has been pointed out above in the description of the mass selection experiment that no "new blood" was introduced into the flock during the course of that experiment. That is to say, no new additions were made to the hereditary constitution of the birds during that period. This certainly represents a condition of some degree of inbreeding, at least, in spite of the fact that an effort was made never to breed close relatives together. As a matter of fact, because of a lack of an adequate system of individual pedigree records during this mass selection experiment, it probably happened on several occasions that quite close relatives were bred together. Whether or not this occurred, it is certain that during 9 years only the "blood"* of relatively few original individuals was represented in the flock, and since all this breeding was in one line, the result can certainly be regarded only as a narrow-bred flock. Whether one chooses to call this "inbreeding" or not depends on his definition of the term. The biological condition and not the term

---

* Meaning *germ-plasm*.

used to designate it is the important thing. That condition was as described.

It amounts to a truism to say that it is one of the strongest and most ancient of breeding traditions that inbreeding is in and of itself harmful, and inevitably results in deterioration. This being so it is clear that a possible interpretation which demands careful attention is that the real reason for the failure of the mass selection experiment to produce increased mean annual production was to be found in the blighting effect of the inbreeding or narrow-breeding (as one chooses) which had been practiced. The deterioration from this cause would be held, on this view, to balance or offset the potential gain assumed to have resulted from the selection.

It seemed very important at the outstart of the new period in the breeding work at this Station, to test carefully the validity of this interpretation. In order to do so the experiment to be described was planned and carried out in 1908 and 1909. The essential point to this experiment was to compare in respect to egg production the offspring of two sorts of matings. In one set of these matings both the male and female mated together were from the Station flock and might be very closely related (even brother and sister). In the other set of matings the female in each case was from the Station flock but the male was a pure-bred Barred Rock cockerel purchased from one or another among the then more or less prominent breeders of this variety in the eastern United States. It will be seen that in the first case the progeny of the matings represent the continuation to the full degree of the narrow breeding practiced during the preceding 9 years. In the second case the progeny represent the widest breeding possible. That is, the male and female are absolutely unrelated and come from entirely distinct strains. If the narrow breeding during the selection experiment was really inimical to high egg production, and brought about deterioration, it would be expected that progeny from parents of absolutely unrelated "blood" would show marked superiority to those from a continuation of the narrow breeding. The maximum effect in the way of rejuvenation from "out-crossing" should show itself here.

In carrying out the experiment 8 Barred Rock cockerels were purchased in the early spring of 1908 from as many different breeders of this variety. These birds will be designated in what follows as "foreign" cockerels. Their sources are indicated in the following list.

Cockerel No.	Source.
56	Mr. C. H. Welles, Stratford, Conn.
57	Pine Top Poultry Farm, Hartwood, N. Y.
58	Gardner & Dunning, Auburn, N. Y.
60	L. J. Bundy & Son, Silver Springs, N. Y.
61	Mr. Geo. E. Mann, Dover, Mass.
65	Mr. Wesley B. Barton, Dalton, Mass.
68	Mr. Geo. W. Hillson, Amenia, N. Y.
70	Mr. M. L. Chapman, Mount View, Farmington, Conn.

All of these males, it may be said, were first-class, vigorous birds.

The breeding in 1908 was done in a house known as No. 1 which has since been destroyed. It was not adapted for use as a breeding house, and the results obtained as to fertility and hatching of eggs were as a consequence poor.* In this house there were 14 pens each accommodating one male and ten females. In addition there were 5 larger pens, each holding one male and 14 females. Four of the small pens were used for cross-breeding work and had no part in the present experiment. In addition to the hens in house No. 1, two breeding pens were mated up in house No. 2 which is of the curtain front, fresh air type, which experience has shown to be much better suited, both to breeding work and to egg production. In each of these two pens one male and 15 females were placed.

The arrangement of the pens relative to the hereditary constitution of the birds is shown in the following scheme.

It will thus be seen that in No. 1 house, pens headed by foreign cockerels and by Station cockerels alternated, whereas both pens in No. 2 house were headed by Station males. For each of the first 7 foreign cockerels there was a Station cockerel working under the same environmental conditions. During the course of the breeding season it was necessary to withdraw a number of the females from each pen and substitute others in their place.

In selecting the females to be bred to these various cockerels it was attempted to make as even a distribution between foreign

---

* Cf. Pearl, R., and Surface, F. M., loc. cit. p. 109.

and Station cockerels as possible. It is essential in such an experiment that the average age, health, vigor, size, and egg-production should be approximately the same in the females mated

House No.	Pen No.	Foreign Cockerels.	Station Cockerels.	Females.
1	5	D70		
	6		D2*	
	7	D60		
	8		D32	
	9	D65		
	10		D5	
	11	D56		
	12		D11	All females from Station stock.
	13	D58*		
	14		D16	
	15	D61		
	16		D35	
	17	D57		
	18		D17	
	19	D68		
2	20		D26	
	21		D31	

* This bird got no adult ♀ progeny in 1908 and is dropped from further discussion here. See text.

to the one class of cockerels as in those mated to the other class. Data regarding this and other matters related to the breeding history of the birds in this experiment are given in Table D. One of the foreign cockerels (No. D58) and one of the Station cockerels (No. D2) produced no adult female

TABLE D.

Data Regarding Breeding History of Individuals in Inbreeding Experiment.

Item.	Foreign Cockerels.	Station Cockerels in House No. 1.	Station Cockerels in House No. 2.
Number of ♂♂ in experiment.	7	6	2
Number of ♀♀ mated with these ♂♂*.	120	93	33
Number of ♀♀ which produced adult daughters.	26	17	15
Percentage of ♀♀ which produced adult daughters.	21.7	18.3	45.5
Mean winter egg production of all ♀♀ mated	33.26	31.96	27.70
Mean winter egg production of all ♀♀ which produced adult daughters.	44.31	39.53	28.27

* This is the *total* number of ♀♀ mated during the whole season and includes individuals put in to replace those removed. The number of ♀♀ with each male in the pen was *at any given time* only the number stated above (p.137).

progeny in 1908 and hence do not appear in the discussion of the fecundity of narrow and broad bred females. On this account they are dropped out of Table D.

From this table a number of points are to be noted.

1. The number of females in this experiment which succeeded in producing adult daughters is relatively small with reference to the total number of females bred. That is to say, there is a rather stringent genetic selection here. Examination of the table shows that this is due to the unfavorable environmental conditions which existed in house No. 1. This was obvious from direct observation during the course of the experiment. The way in which it is shown in the table is that whereas but 21.7 per cent and 18.3 percent respectively of the females mated with foreign and Station cockerels in No. 1 house got adult female progeny, on the other hand 45.5 percent of the females mated in house No. 2 got adult daughters. This record of 45.5 per cent for the females in house No. 2 is itself lower than it would have been had not one of the males used in No. 2 house (No. D26) proved to be a rather unsatisfactory bird, which did not make a very good record either in respect to fertility or hatching quality of eggs.

2. The rather stringent genetic selection brought about by the unfavorable conditions in house No. 1 does not interfere with the experiment in regard to the effect of inbreeding because the force of this selection was substantially the same in the case of females mated in No. 1 house with foreign cockerels and females mated in the same house with Station cockerels. The difference is only 3.4 percent and cannot be regarded as significant.

3. With respect to the mean winter egg production of the mated females it is seen that the averages for the females which were originally mated with foreign cockerels, and that for the females in house No. 1 originally mated with Station cockerels are substantially the same. What advantage there is is in favor of the foreign cockerels, that is, of the broad-breeding side of the experiment. The probable errors of these averages of egg production lie in the neighborhood of 1.2. It is obvious that the difference (1.30) between the mean production of the two sets of females in No. 1 house is only such as might be expected to arise from random sampling. The females mated with Sta-

TABLE E.

Frequency Distributions of Egg Production of Daughters from Out-cross and Narrow Matings.

NUMBER OF EGGS LAID	WINTER PERIOD.				SPRING PERIOD.			
	Daughters of foreign ♂♂ (out-cross)	Daughters of Station ♂♂ in house No. 1 (narrow-bred)	Daughters of Station ♂♂ in house No. 2 (narrow-bred)	Daughters of *all* Station ♂♂ (narrow-bred)	Daughters of foreign ♂♂ (out-cross)	Daughters of Station ♂♂ in house No. 1 (narrow-bred)	Daughters of Station ♂♂ in house No. 2 (narrow-bred)	Daughters of *all* Station ♂♂ (narrow-bred)
0–4	20	17	30	47	4	2	1	3
5–9	10	14	11	25	1	2	—	2
10–14	17	19	15	34	1	4	1	5
15–19	16	12	7	19	1	1	1	2
20–24	14	14	13	27	3	3	7	10
25–29	21	8	9	17	5	7	5	12
30–34	15	10	22	32	11	11	9	20
35–39	10	8	7	15	13	6	12	18
40–44	7	5	7	12	24	6	18	24
45–49	6	5	5	10	21	13	14	27
50–54	5	6	3	9	12	14	13	27
55–59	4	3	4	7	13	17	21	38
60–64	2	3	4	7	21	16	14	30
65–69	5	2	2	4	5	7	11	18
70–74	3	1	5	6	5	4	3	7
75–79	2	2	3	5	1	2	1	3
80–84	1	—	—	—	—	—	—	—
85–89	1	—	—	—	—	—	—	—
90–94	1	—	—	—	—	—	—	—
Totals	159	129	147	276	141	115	131	246

tion cockerels in house No. 2 have a somewhat lower mean winter production than those in No. 1 house. These were the last two pens to be mated and it was then not possible any longer to select as many high layers to offset (in the average) the low layers as had been done in mating up the pens in No. 1 house.

4. Considering next the females which produced adult daughters, the differences in winter production are much more marked than when all mated females are taken together. The means in this case, however, stand in the same relation to each other as when all females are included. As would be expected the mean production of the *mothers* of adult daughters is higher than the mean production of all females bred in the same class. This merely means that the better a hen lays the more likely she is to have adult progeny, because she has more chances. The mothers of adult daughters of foreign cockerels have a mean winter production more than 16 eggs higher than that of the mothers of daughters of the two Station males in house No. 2. The difference is much smaller between the two sets of mothers in house No. 1 though there it is probably statistically significant. The females in the out-cross matings which produced adult daughters thus have the higher egg production records. If this factor has any influence at all it would evidently act in accord with any beneficial result of the outcrossing itself to help to produce higher laying in the progeny.

Let us turn next to the results of the experiment. The frequency distributions for the winter * and spring (March 1 to June 1) egg production of the daughters of (a) foreign cockerels, (b) Station cockerels in house No. 1 and (c) Station cockerels in house No. 2 are given in Table E.

The usual biometric constants for these distributions are presented in the following table.

The means of Table F are shown graphically in fig. 77.

---

* Cf. p. 154 *infra*.

TABLE F.

Constants of Variation in Egg Production of Progeny of Out-cross and Narrow Matings.

CONSTANT.	Winter Period.				Spring Period.			
	Daughters of foreign ♂♂ (out-cross)	Daughters of Station ♂♂ in house No. 1 (narrow-bred)	Daughters of Station ♂♂ in house No. 2 (narrow-bred)	Daughters of *all* Station ♂♂ (narrow-bred)	Daughters of foreign ♂♂ (out-cross)	Daughters of Station ♂♂ in house No. 1 (narrow-bred)	Daughters of Station ♂♂ in house No. 2 (narrow-bred)	Daughters of *all* Station ♂♂ (narrow-bred)
Mean (or average)	28.03 ± 1.07	25.14 ± 1.12	26.45 ± 1.16	25.83 ± 0.81	46.68 ± 0.86	47.24 ± 1.08	48.11 ± 0.84	47.70 ± 0.68
Standard deviation	19.92 ± 0.75	18.85 ± 0.79	20.77 ± 0.82	19.91 ± 0.57	15.14 ± 0.61	17.12 ± 0.76	14.33 ± 0.60	15.70 ± 0.48
Coefficient of variation	71.0 ± 3.8	75.0 ± 4.6	78.5 ± 4.6	77.1 ± 3.3	32.4 ± 1.4	36.2 ± 1.8	29.8 ± 1.3	32.9 ± 1.1

From these data it appears that there is no substantial difference in egg production, either in the winter or the spring periods, between the pullets which came from entirely unrelated parents on the one hand, and those whose parents both belonged to the Station strain on the other hand. In the winter period

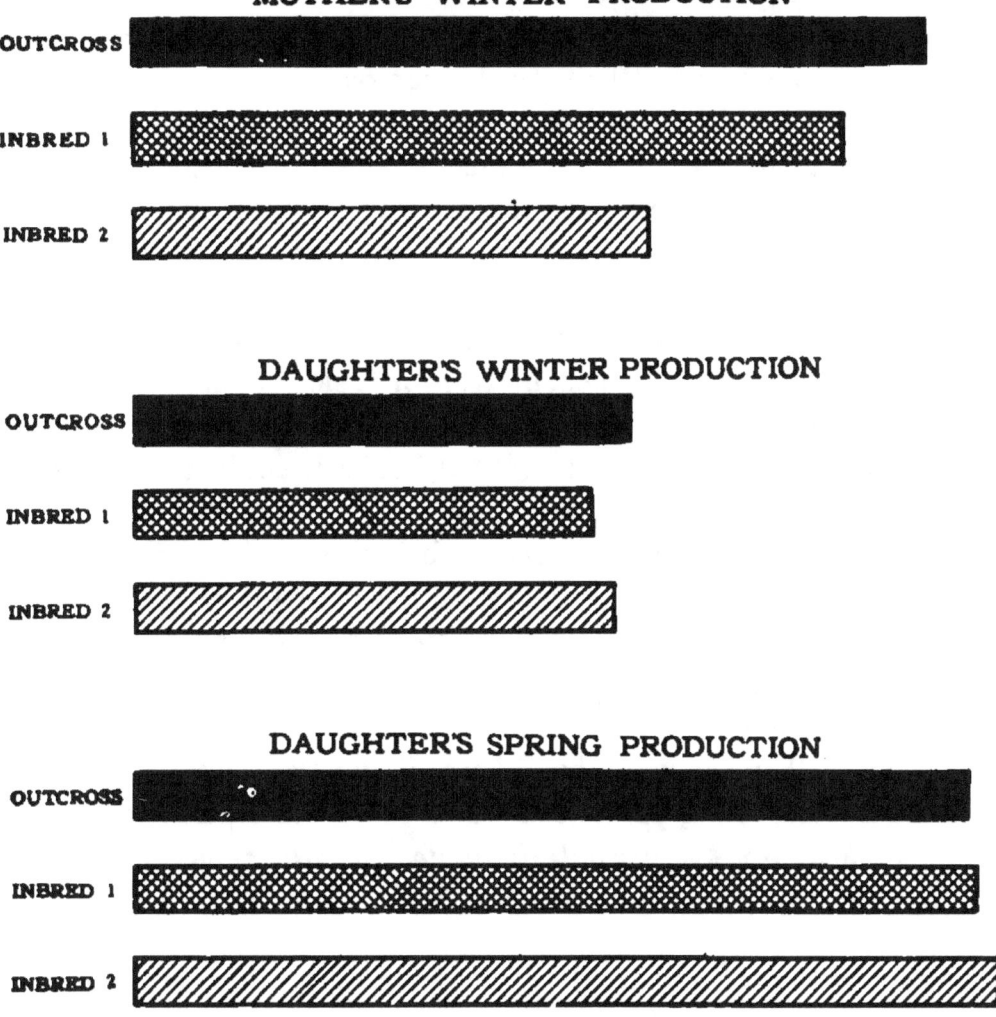

Fig. 77. Diagram showing the mean winter and spring egg production of the daughters of outcross and narrow matings.

it is true that the birds from out-cross matings have a slightly higher mean production than either group of narrow-bred daughters. The difference, however, is too small to be significant. Thus the greatest difference in the table is between the winter means for the out-cross matings and the narrow matings in No. 1 house. Here, however, the difference is but 2.89 ± 1.55, an amount less than twice its probable error. In the spring period the daughters from out-cross matings actually laid a little less than those from the narrow matings.

An examination of adult mortality records leads to the same conclusion as the fecundity records. The percentage mortality amongst the adult daughters from out-cross matings was 11.3 per cent. Among the daughters from narrow matings in house No. 1, the mortality was 10.8 per cent. For the daughters of *all* narrow matings the mortality rate was the same, viz. 10.8. Thus it is clear that as measured by this index the out-crossed stock was no more vigorous than the narrow bred.

It may further be said that not only was there no difference in the first generation between the offspring of out-cross and narrow matings, but further there was no difference in the progeny in subsequent generations. Putting in "new blood" did not improve or rejuvenate the stock. One must then conclude either (a) that the stock was so hopelessly and completely degenerate as to be past any benefit from infusion of new "blood," or (b) that the stock was *not* deteriorated in respect of constitutional vigor or vitality and that therefore it could not be expected that out-crossing would have any rejuvenating effect. The first conclusion certainly cannot be deemed the correct one in view of the evidence presented in this bulletin in respect to mortality, egg production, etc. The stock has never been in the condition of excessively low constitutional vigor which would be demanded by such a conclusion.

It would then appear that *there is no evidence that the amount of inbreeding practiced during the mass selection experiment had any unfavorable influence on either the egg production or the general vitality of the stock.*

PART III.

NEW PLAN OF BREEDING FOR EGG PRODUCTION.

At the end of the period from 1898 to 1907 during which the mass selection experiment had been carried on it seemed advisable to inaugurate a change in the plan of investigation. Considering the results obtained, and the fact that a large amount of statistical data had been accumulated and was available for analysis, it appeared unlikely that further continuation of mass selection would yield results of sufficient value to warrant carrying on the work. The ultimate object of the work at the Maine Station in this field was, and is, to get at the underlying principles of the inheritance of fecundity in fowls. One set of facts having been accumulated, efforts were turned towards getting more data of a somewhat different character.

The keynote to the new line of investigation has been the analysis of the inheritance of fecundity by means of *individual pedigrees*. By this method one determines precisely the behavior of each individual in inheritance. Those individuals of like hereditary behavior or performance may then be lumped together for statistical treatment if desired. The "individual pedigree" is the nearest approach which can be made in an organism in which each individual is of one sex only to the *genealogical* unit termed by Johannsen a "pure line" in self-fertilizing plants. Its employment in the analysis of inheritance in animals has underlying it the same considerations which make the "pure line" so potent an instrument of research in plants and non-sexually reproducing animals.

In order that what follows in Part IV may be more readily understood it is desirable here to explain fully the working hypothesis which the present study in the inheritance of fecundity is testing. To put the matter most briefly it may be said that this hypothesis is an adaptation to the particular case in hand of the genotype concept of Johannsen. In more detail the case is as follows.

Johannsen in his work on beans* brought out very clearly three things which in themselves and in their implications are of fundamental importance to all practical breeders of animals or plants, as well as to students of breeding. These three things are:

1. That the size of an individual bean was no absolute or certain criterion whatever as to the average size of its offspring. He found that while some particular large beans always produced large offspring beans, other equally large ones always produced small offspring beans. Some individual small beans produced offspring of large average size, others produced beans of small average size like the parent, and, in general, he showed it to be quite impossible for anyone to tell merely from the size of a bean itself whether its progeny will be large or small.

---

*Johannsen, W., Ueber Erblichkeit in Populationen und in reinen Linien, Jena, 1903.

The nature of Johannsen's results on this point have been very cleverly set forth in the accompanying diagram which is taken from a paper by Wood and Punnett.*

2. That a population of beans, no matter from how supposedly "pure" a commercial variety it is taken, is really not a homogeneous unitary aggregation, but instead is made up of a varying number of lines or strains, each of which breeds true to itself when propagated in isolation. In other words the popu-

	SMALL SEED		AVERAGE SIZED SEED			LARGE SEED	
	1	2	1	2	3	1	2
Single Seeds Picked out of Trade Samples							
Average Size of Seed on Plant grown from above Seed							
Average Size of Seed on plant grown from a Seed of the Above Plant							

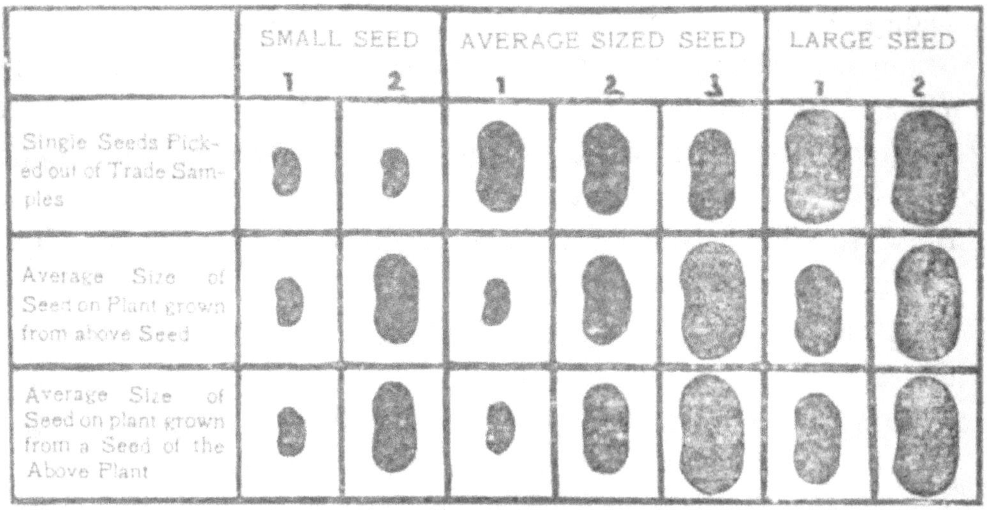

Fig. 78. Diagram to illustrate Johannsen's results with beans. (From Wood and Punnett).

lation in question is a mixture of several component lines. The individuals in each line produce offspring true to the type of the line, rather than to the type of the population as a whole, except in cases where by chance the population type and the type of one or more lines happen to be the same.

3. That when mass selection alters the population type it does so by a process of isolating from the mixture certain strains whose own types are different from the original general population type, and which differ in the direction towards which selection was made. Thus if one begins in a general mixed population of beans to select for planting the largest beans, and by so doing increases the average size of bean in the crop, what he really does is gradually to throw away all beans except those

---

* Wood, T. B., and Punnett, R. C., Heredity in Plants and Animals. Mendel's Principles and their Bearing on Agricultural Problems. Trans. Highland and Agr. Soc. Scotland. Ser. 5, Vol. XX. pp. 36-86, 1908.

which belong to strains having large beans as the type. Having isolated from the population one of these component strains which breeds true to a definite type no amount of further selection will modify that strain. In other words Johannsen showed that, in beans at least, selection is only effective to isolate or pick out what heritable variations were already present as components of the population to begin with. Selection within a line or strain is ineffective.

These and other results of recent work (particularly that along Mendelian lines) lead to a new conception of the mechanism of heredity which differs markedly from older views. The keynote to this conception is that it is the germ cell (egg or sperm) and not the body or soma which is the factor of primary importance in inheritance. What the individual is like in respect to its personal, somatic * characters is not determined by the somatic characters of its parents, but by the composition or constitution of the parental gametes. Thus the size of a bean is determined not by the *size* of its parent bean, but by the gametic constitution of the latter.

Experimental breeding along Mendelian lines has shown very clearly that many characters of organisms are inherited as separate units, so that by proper cross-breeding new combinations of characters may be made. Thus, for example, suppose one crosses together a Barred Rock, which is a barred bird with a single comb, and a Cornish Indian Game, which is a non-barred bird with a pea comb. In the second generation he will have barred birds with pea combs, barred birds with single combs, non-barred birds with pea combs and non-barred birds with single combs occurring in certain definite proportions to each other. This result shows beyond question that, whatever the mechanism, comb form is inherited separately from plumage pattern. These characters behave as separate units.

---

* For the reader not familiar with the technical terminology of biology, it may be said that "somatic" is used in designation of those characters of the organism which pertain to all parts except the reproductive or germ-cells. These reproductive cells are called "gametes." We then have the adjective "gametic," meaning "pertaining to the germ cells," in contrast to "somatic" meaning "pertaining to any or all parts of the organism *other than* the germ cells."

Each character of an organism which is inherited must, in some manner, be represented by a factor in the germ cell. For each of these gametic factors Johannsen proposes the name *gene*. A gene then is that factor in the germ cell or gamete the presence of which (and therefore its taking part in gametic reactions) is connected with the existence of a particular somatic character or set of characters. All of these unit factors or genes taken together constitute the *genotype* of the organism. The *genotype* then represents the hereditary or gametic constitution of the individual as distinguished from the somatic.

Johannsen's experiments show that in the organisms with which he has worked genotypes cannot be modified by selection. That is, given a group of a thousand individuals all of the same genotypical constitution, and no amount of selection of somatic variations within this group will produce any permanent or inherited effect. The offspring of the selected individuals will be, on the average, like the offspring of other individuals which had the same genotypical constitution. The sole function of selection then becomes the *isolation* and propagation of strains composed of individuals having the desired genotypical constitution.

It may be said that confirmation of one or another of the essential features of the genotype concept of Johannsen, has come not only from his work with beans and barley, but also from the work of Nilsson-Ehle, Jennings, Shull, East, Hanel, Roemer and others on a variety of plants and animals.

Suppose this general conception of the mechanism of inheritance to be taken as a working hypothesis (to be tested by experiment) in attempting to increase or decrease egg production in the domestic fowl by breeding. To what sort of picture of the make-up of a flock does it lead? First of all it may be assumed that a number of distinctly different genotypical constitutions will be represented in the flock. By way of concrete illustration let us suppose a population or flock to be made up of individuals representing seven marked genotypical differences. Each set of individuals of like genotypical constitution may be considered to form a "line" or "strain" in the breeder's sense. There will then be seven lines which may be designated as $A, B, C, D, E, F,$ and G. (See fig. 79). The general average production for the population as a whole is 130. The geno-

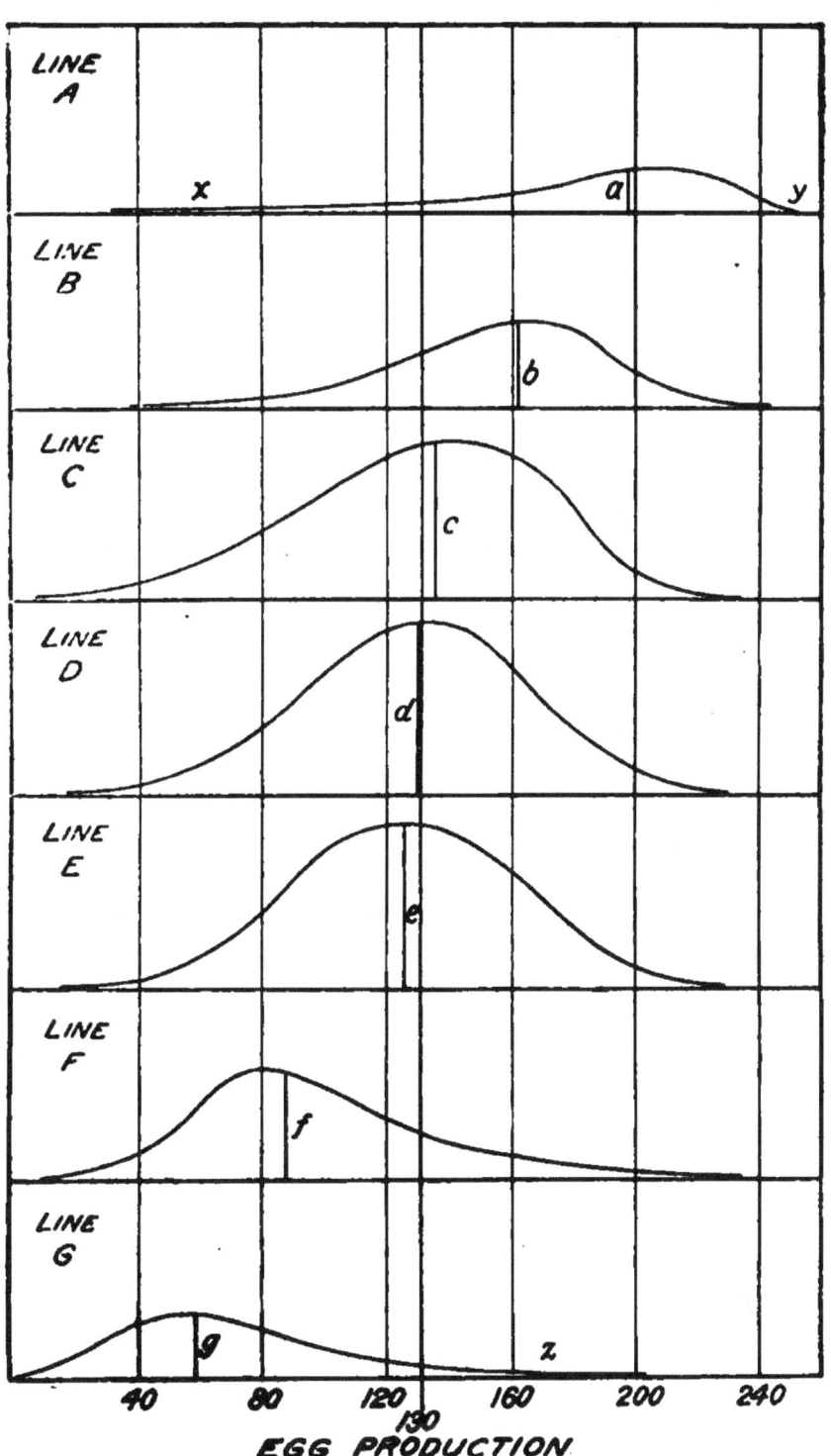

Fig. 79. Diagram to illustrate hypothesis regarding inheritance of egg production.

types of the several lines, are however, quite different. We may consider these genotypes to be given by the small letters, a, b, c, d, e, f, g. It is seen that three of the lines have genotypes corresponding closely to the general flock average 130. Two genotypes are well above the general average of the flock, and two others below. Now the basic idea of the pure-line concept as here applied is that if, for example, two individuals of line A are bred together, the average productivity of the offspring will be approximately a (say 197 eggs) regardless of the egg record of the particular female bird bred. That is, it is to be expected that the average productivity of the progeny of a female of line A with an egg record at x (say 80 eggs) and of a female with a record at y (say 240 eggs) *will be the same* provided both are mated to male birds from the same line as themselves, that is, the A line.

The same idea may be illustrated in another way. It is to be expected that in such a population as that illustrated in the diagram the average egg production of the daughters of 200-egg hens will be as indicated in the following table:

If a "200-egg" hen comes from—	The probable average production of her daughters will be about—
The A line, and is mated with an A line male....	197
The B line, and is mated with a B line male....	162
The C line, and is mated with an C line male....	134
The D line, and is mated with a D line male....	130
The E line, and is mated with a E line male....	125
The F line, and is mated with an F line male....	88
The G line, and is mated with a G line male....	58

Similarly we have according to the pure-line concept the following expectation regarding the progeny of poor producers:

If a "75-egg" hen comes from—	The probable average production of her daughters will be about—
The A line, and is mated with an A line male....	197
The B line, and is mated with a B line male....	162
The C line, and is mated with a C line male....	134
The D line, and is mated with a D line male....	130
The E line, and is mated with an E line male....	125
The F line, and is mated with an F line male....	88
The G line, and is mated with a G line male....	58

In other words it would be expected on the hypothesis set forth, and in so far hypothesis and actual fact are in precise accord, that the record of egg production in and of itself alone is not a criterion of particular signficance in selecting females for breeding for improved egg production in the flock. It appears to be of vastly more importance to know the genotype of the line to which the individual belongs. This idea is already perfectly familiar to the successful breeders of poultry for fancy points. What is here called "blood line" they usually call a "strain."

In 1907-08 a plan of breeding designed to test the general hypothesis above set forth was put into operation. A statement of a portion of the results which have been obtained to date is given in Part IV. This part of the bulletin is a reprint of an address given before the American Society of Naturalists in December, 1910. It seems advisable to reprint it in entirety here, though to do so involves some repetition of points already brought out in this bulletin. For this the author would ask the reader's pardon.

PART IV.

INHERITANCE OF FECUNDITY IN THE DOMESTIC FOWL.*

There are under discussion at the present time two general views regarding certain fundamental points in heredity. Each of these points of view has its zealous adherents. On the one hand is what may be designated the "statistical" concept of inheritance, and on the other hand, the concept of genotypes. By the "statistical" concept of inheritance is meant that point of view which assumes, either by direct assertion or by implication, that all variations are of equal hereditary significance and consequently may be treated *statistically* as a homogeneous mass,

---

*Papers from the Biological Laboratory of the Maine Experiment Station, No. 25. This paper was read at the meeting of the American Society of Naturalists at Ithaca, December, 1910. It was first published in the American Naturalist, Vol. XLV, pp. 321-345, June, 1911, and is here reprinted without change except that the numbers of the figures are here changed to accord with the preceding figures in this bulletin. Figures 1 to 5 inclusive of the original publication become figs. 80 to 84 respectively in this reprint.

provided only that they conform to purely statistical canons of homogeneity. This assumption of equal hereditary significance for all variations is tacitly made in deducing the law of ancestral inheritance, when individuals are lumped together in a gross correlation table.* The genotype concept, on the other hand, takes as a fundamental postulate, firmly grounded on the basis of breeding experience, that two sorts of variations can be distinguished, namely those (a) that are represented in the germinal material and are inherited without substantial modification, as in "pure lines," and those (b) that are somatic and are not inherited. By anything short of the actual breeding test it is quite impossible to tell whether a particular variation observed in the soma belongs to the one category or to the other. As I have tried to emphasize in other places, it is both to be expected on this view of inheritance, and is also the case in actual fact, that the somatic manifestation or condition of any character is a most uncertain and unreliable criterion of the behavior of that character in breeding. Finally under the genotype concept, of course, the whole array of facts brought out by Mendelian experiments find their place.

Now while certain adumbrations of the genotype concept have long been current in biological speculations in regard to heredity, this general view-point owes its grounding in solid facts primarily to Johannsen's work with beans and with barley. It is to be noted that in these cases, as well as in most of the investigations of the pure line theory which have followed Johannsen's work, the organisms used have been such as reproduced either by self-fertilization, or by fission, or by some vegetative process. This brings us to the consideration of a question of great importance, both theoretical and practical. In cases of diœcious organisms, where a *"pure"* pedigree line in the sense that such lines are found in beans or in Paramecium by definition can not exist, has the genotype concept any bearing or significance? In a general way it obviously has. Probably no one (except possibly some of the ultra-statistical school) could be found who would deny that in general a distinction is to be made between variations having a gametic and those hav-

---

* For a more detailed discussion of this point see a paper by the present writer entitled "Biometric Ideas and Methods in Biology; their Significance and Limitations," in the *Revista di Scienza* (in press).

ing merely a somatic basis. But specifically how far has the genotype concept any application in case of "non-selfed" organisms? Johannsen in his "Elemente" has thoroughly analyzed Galton's material and shown that it is capable of a satisfactory and reasonable interpretation on the genotype hypothesis, and East and Shull have gone far in the analysis of genotypes in maize. This, however, is only a beginning. There is the greatest need for careful, thorough investigations of the inheritance of characters showing marked fluctuating variation in organisms having the sexes separate. Here lies one of the crucial fields in the study of inheritance to-day. Through the brilliant results in Mendelian directions and from the study of really *"pure"* lines we are getting clear-cut ideas as to the inheritance of qualitatively differentiated characters, such as color, pattern and the like, on the one hand, and in regard to the inheritance of quantitative variation in self-fertilized or non-sexually reproducing organisms, on the other hand. But beyond all these lie the difficult cases where in dioecious forms quantitative variations must be dealt with. If these can be cleared up and brought harmoniously into a general scheme or view-point regarding inheritance, we shall have gone a long way in the solution of this world-old biological problem.

For some four years past the writer has been engaged in a study of the inheritance of fecundity in the domestic fowl. The problem presented here is an important one from the practical as well as the theoretical standpoint. If definite and sure methods of improving the average egg production of poultry by breeding can be discovered it will mean much to the farmers of the nation. At the same time egg production is a character admirably adapted to furnish definite and crucial data regarding inheritance. Variations in egg production are readily measured, and can be directly expressed in figures.

The general results of this study of the inheritance of fecundity may be said, in a word, to be so far as they go in entire accord with the genotype concept, and not to agree at all with the "statistico-ancestral" theory of inheritance. Indeed so ill is the accord here that the chief exponent of the latter doctrine has recently attempted to throw the whole case out of court * by

---

* Pearson, K., "Darwinism, Biometry and some Recent Biology, I," *Biometrika*, Vol. 7, pp. 368-385, 1910.

asserting that fecundity is not inherited in fowls, and that the present writer's investigations show essentially nothing more than that. It will be the purpose of this paper to present some figures sufficient to indicate with some degree of probability, I think, first that egg production in fowls *is* inherited, and second that it is probably inherited in accord with the genotype concept, in spite of the fact that we do not and can not here have "pure lines" in the strict sense of Johannsen's definition. In the present paper, owing to limitations of space, the whole of the data in hand obviously can not be presented. Only a few illustrative cases can be given here.

Before entering upon the discussion of the evidence it is necessary to call attention to two points. The first is in regard to the unit of measuring egg production used in the work. For reasons which have been discussed in detail elsewhere * the unit of study has been taken as the egg production of the bird before March 1 of her pullet year. This "winter production" is a better unit for the study of the inheritance of fecundity than any other which can be used practically. All records of production given in this paper are then to be understood as "winter" records, comprising all eggs laid up to March 1 of the first year of a bird's life. It may be said that the "normal" mean winter production of Barred Plymouth Rocks (the breed used in this work) is fairly indicated by the 8-year average of the Maine Station flock. This average November 1 to March 1 production is 36.12 eggs.* This figure is based on eight years continuous trapnesting of the flock with which the present work was done, carried out before these investigations were begun.

In the second place it is desirable to call attention to some of the difficulties which attend an attempt to analyze the inheritance of the character egg production. The most important of these is the fact that this character is not visibly or somatically expressed in the male. A male bird may carry the genes of high

---

* Bull. Me. Agr. Exp. Sta., No. 165. U. S. Dept. Agr. Bur. Anim. Ind., Bul. 110, Part II.

* It should be said that up to and including the winter of 1907 only the November 1 to March 1 records are available as a "winter" record. Since that time the small number of eggs laid before November 1 (on the average two or three per bird) are included in the "winter" totals. These, then, give, as stated, the total production up to March 1.

fecundity, but the only way to tell whether or not this is so is to breed and rear daughters from him. All Mendelian workers will agree that it is sometimes difficult enough to unravel gametic complexities in the case of characters expressed somatically. It is vastly more difficult when only one sex visibly bears the character. In the second place a very considerable practical difficulty arises from the fact that egg production is influenced markedly by a whole series of environmental circumstances. The greatest of care is always necessary, if one is to get reliable results, to insure that all birds shall be kept under uniform and good conditions. Further, on this account, it is necessary to deal with relatively large numbers of birds. Some of the important conditions to be observed in work on fecundity have been discussed elsewhere * and need not be repeated here.

Turning now to the results we may consider first

THE EFFECT OF SELECTION FOR FECUNDITY IN THE GENERAL POPULATION.

On the "statistico-ancestral" view of inheritance it would be expected that if fecundity were inherited at all this character would respond to continued selection. That is, it would be expected, if the highest layers only were bred from in each generation, that the general flock average would steadily, if perhaps slowly, increase and that any level reached would be at least maintained by continued selection. In 1898 an experiment in selecting for high egg production was begun at the Maine Station. In this experiment only such females were used as breeders as had laid over 150 eggs in their pullet year (corresponding roughly to an average winter production of 45 or more eggs) and the only males used were such as were out of birds laying 200 or more eggs in the year. This experiment was continued until the end of 1908. The selection, be it understood, was based on the egg record alone, and no account was kept of pedigrees or of genotypes. Every female with a record higher than 150 eggs in the year was used as a breeder regardless of whether her high fecundity was genotypic or phænotypic.

The results of this selection experiment covering a period of nine years have been fully reported elsewhere.* Here it needs

---

* Me. Agr. Exp. Sta. Ann. Rept. for 1910, p. 100.
* U. S. Dept. Agr. Bur. Anim. Ind., Bul. 110, Parts I and II, 1909 and 1911. *Zeitschr. f. indukt. Abst. a. Vererb.-Lehre*, Bd. 2, 1909, pp. 257-275.

only to be said that the net outcome of the experiment was to show that there was no *steady* or *fixed* improvement in average flock production after the long period of selection. There was no *permanently* cumulative effect of the eight (in the last year) generations of selected ancestry. So far from there having been an increase there was actually a decline in mean egg production concurrent with the selection, taking the period as a whole. During parts of the selection period, however, as for example the years 1899-1900 to 1901-'02, inclusive, and the years 1902-'03 to 1905-'06, inclusive, an improvement from year to year was to be noted, but in each case the flock dropped back in intervening years. This is an important point, the meaning of which is now clear. The flock average from year to year depended largely upon *whether the breeders of the year before had had their high fecundity genetically represented or only somatically*. In some years the selection was fortunate in getting nearly all the breeders from good (*i. e.*, "high production") genotypes or from good *combinations* of genes. In other years just the opposite thing happened; the high layers chosen as breeders came from low genotypes or combinations of genes. The general upshot was that while the selection of *high layers* merely as such was systematic year after year the result attained in the general flock production was entirely haphazard and uncertain. This is exactly what would be expected on the genotype hypothesis, but not on the "statistico-ancestral."

TABLE I.

MEAN WINTER (NOVEMBER 1 TO MARCH 1) EGG PRODUCTION DURING THE SELECTION EXPERIMENT.

Year	Mean Winter Production
1899-1900	41.03
1900-01	37.88
1901-02	45.23
1902-03	26.01
1903-04	26.55
1904-05	35.04
1905-06	40.66
1906-07	21.44
1907-08	15.92

The actual course of the average winter egg production (not hitherto published) during the period is given by the figures of Table I and shown graphically in fig. 80.

Certainly the first line of evidence, derived from a long-continued experiment, involving more than 2,000 individuals, gives no support to the "statistico-ancestral" theory and indeed is in flat contradiction to one of the most fundamental tenets of that faith.

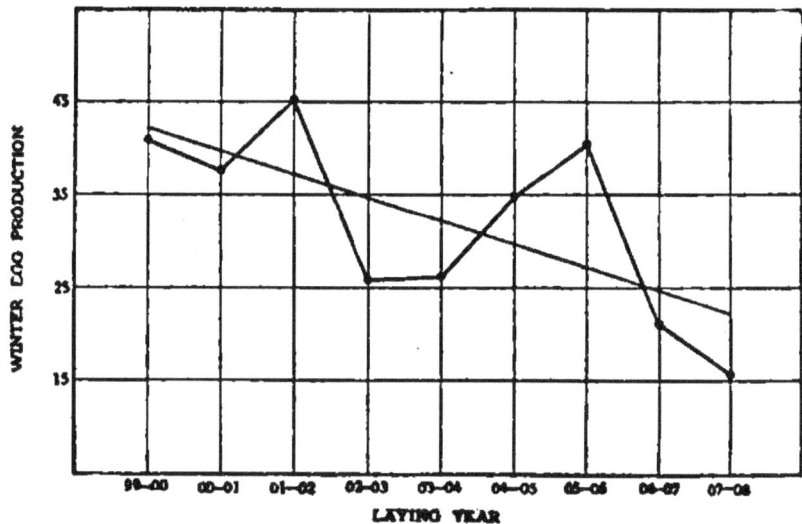

FIG. 80. Diagram showing the course of average winter egg production during the period covered by the mass selection experiment.

Let us next consider the question,

ARE SOMATICALLY EQUAL VARIATIONS IN FECUNDITY OF EQUAL HEREDITARY SIGNIFICANCE?

In the spring and summer of 1907 were reared 250 pullets, all of which were the daughters of hens that had laid approximately 200 or more eggs in the first year of their life. This group of mothers was reasonably homogeneous in respect to records of egg production. All had laid about the same number of eggs. Their daughters were, however, far from a homogeneous lot with respect to egg production.* It is plain from the results obtained in that experiment that the egg record of a hen is a most unreliable criterion of the probable number of eggs which her daughters will lay. This is demonstrated by examination of individual cases. Thus consider the two mothers Nos. 253 and 14. Their winter production records were nearly identical (65 and 66 eggs, respectively). Their daughters' aver-

---

* Full details regarding this experiment have been published as Bulletin 166, Me. Agr. Exp. Sta., 1909. See particularly Table I.

age winter productions were 23.87 and 2.40 eggs, respectively! Certainly it seems reasonable to conclude that the gametic constitutions involved in the breeding of 253 and 14 were quite different, though both these hens laid the same number of eggs. Again, take birds No. 386 and 911. One had a winter record of 55 and the other of 52 eggs. Yet their daughters' average winter productions were, respectively, 4.88 and 27.33 eggs. Many more instances of this kind could be brought forward. Taken together, the whole evidence shows beyond the shadow of a doubt that the presence of high fecundity in an individual, and that factor which makes high fecundity appear in the progeny, are two very different things, either of which may be present in an individual without the other. We plainly have here the basis for the distinction of phænotypes and genotypes just as in beans.

THE INHERITANCE OF EGG PRODUCTION IN PEDIGREE LINES.

Let us now consider some of the evidence that such things as genotypes of fecundity really exist in fowls. We may first examine some representative pedigrees covering four generations

Pedigree Line D5D39

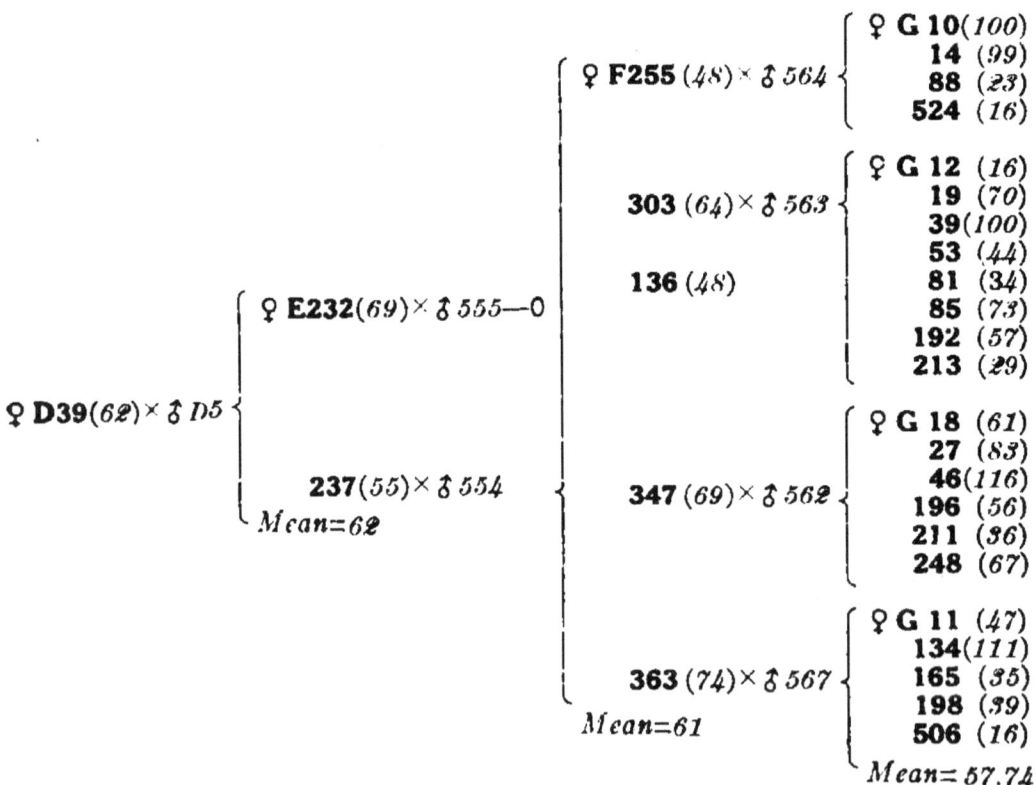

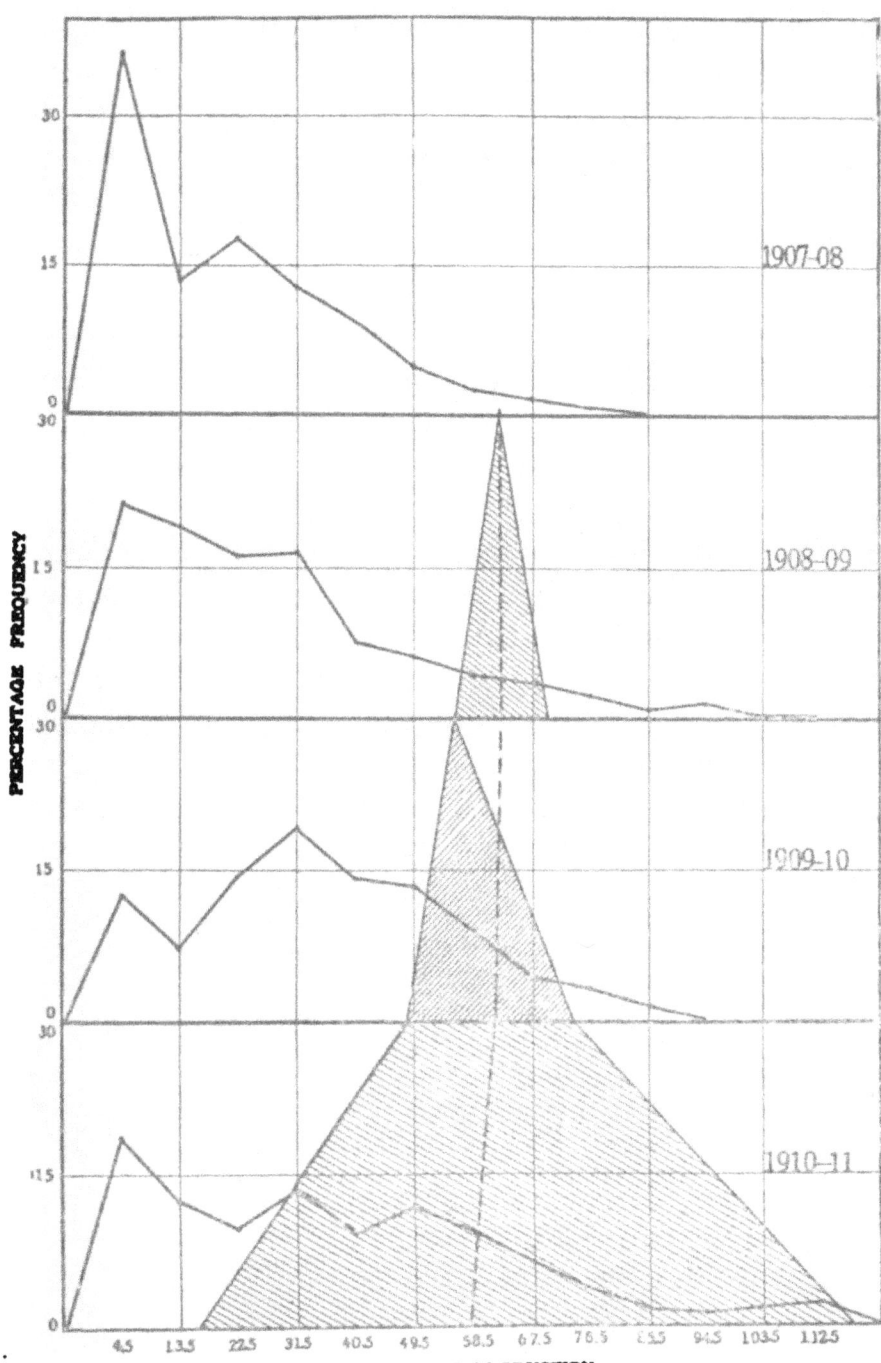

FIG. 81. Diagram showing range of variation and mean fecundity in each generation of line D5D39. The main polygons of variation give the distribution of fecundity in the general flock in each generation. The cross-hatched areas represent the pedigree line, and the heavy dotted lines through these areas represent the mean fecundity of the line in each generation. [It should be understood that in this and the following line diagrams the cross-hatched areas are *not* frequency polygons. They merely give (a) the upper and lower ends of the *range* of variation in the line in each generation, (b) the mean of the line, and (c) the portion of the range which was bred from in each generation to produce the next generation. Note added in reprinting.]

and showing the occurrence of high and low fecundity lines.

As a typical example of a high fecundity pedigree line in which the high fecundity is genotypic, line D_5D_{39} may be considered. In the presentation of this and other pedigree tables the following conventions are adopted. The band numbers of the birds are in bold-faced type, and following the band number of each female, her winter egg record is given in italic figures enclosed in parenthesis. The band numbers of males are given in italics.

This line is shown graphically in Fig. 81.

Little comment on this pedigree line is necessary. We see a certain high degree of fecundity faithfully reproduced generation after generation. Different males were used with different females, but in every case the males used were from high fecundity lines and were believed to carry this quality in their germ cells either in homozygote or heterozygote condition.

In marked contrast to the last example let us consider the *low* fecundity lines $D_{61}D_{168}$. It is a troublesome matter to propagate the low fecundity lines, because of the difficulty of getting a sufficient number of eggs during the early part of the breeding season. The line $D_{61}D_{168}$ is of interest not alone as an illustration of a typical low line, but also because there appeared in it a mutation, or something very like one. We will consider here only the main line and not the mutant.

Pedigree Line $D_{61}D_{168}$

♀**D168**(*33*) · ♂*D61* {
　♀**E231** (*25*) × ♂*552*　　♀**F233**(*32*) × ♂*573*—0
　　　　　　　　　　　　　　　　　　　　　　{ ♀**G221**(*16*)
　419 (*9*) × ♂*551*　　♀**F165**(*7*) × ♂*569* { **430**(*12*)
　　　　　　　　　　　　　　　　　　　　　　{ **477** *1*
　209 (*38*) × ♂*555*—0　　　　　　　　　Mean=9.67
　313 (*26*) × ♂*554*
　363 (*11*) × ♂*550*　　　**174**(*21*)
　15 (*18*)　　　　　　　　♀**F249**(*30*)
　163 (*9*)　　　　　　　　Mean=22
　200 (*12*)
　141 (*0*)
　116 (*28*)
　151 (*11*)
　24 (*23*)
　Mean=17.5
　♀**E248** (**48**)*
}

---

*This was the mutant referred to. Its progeny will be considered later. See p. 161.

This line is shown graphically in Fig. 83, in which the mutant and its progeny are also shown.

A low line in which no mutant has appeared, but in which also the mean production is not so low as in line D61D168 is D65D366. Since the egg production has not been so low in the

Pedigree Line D65D366

```
                                    ⎧ ♀F309(0D)*
                  ♀E239(24)×♂553  ⎨   263(44)           ⎧ ♀G  34  (4)
                                    ⎨   362(43)              42 (37)
                                    ⎩   216(41)×♂569 ⎨      56 (40)
                                                              164 (5)

                  224(43)×♂554    ⎧ ♀F301 (7)
                                    ⎨   223(14)
                                    ⎩   221(42)
♀D366(33)×♂D65 ⎨
                                                            ⎧ ♀G 65(28)
                  354(15)×♂551   ⎧ ♀F242(21)              209(33)
                                    ⎩   221(39)×♂566  ⎨    267(25)
                                                              502(21)
                                                              544 (8)

                  331(31)×♂552-0
                  344(17) ⎧ ×♂550  ⎧ ♀F271(37)
                           ⎩ ×♂528  ⎨ ♀F171(46)

  Mean=26                  Mean=33.4            Mean=22.33
```

early part of the breeding season with this line it has been easier to propagate it.

This line is shown graphically in Fig. 82.

In the examples thus far given we have had to do with pedigree lines in which a given degree of fecundity reappeared from generation to generation with practically no change. In two instances quite certainly, and possibly in several others, a new and distinct variation has suddenly appeared within a line and thereafter bred true, thus presenting the characteristic phenomena of mutation. The most striking instance of this sort occurred in line D61D168 and may be given here in detail. The main part of this line has already been discussed (p. 160). It will be recalled that it is a line of low fecundity. In 1908 there appeared in it one individual of distinctly higher fecundity than any other bird in the large family of that year. This individual when bred produced only high layers. In the next generation two of these daughters were bred to males known to belong to high fecundity genotypes (♂ ♂ 554 and 566). One of these matings unfortunately produced no adult female offspring. The

* Bird died during winter period.

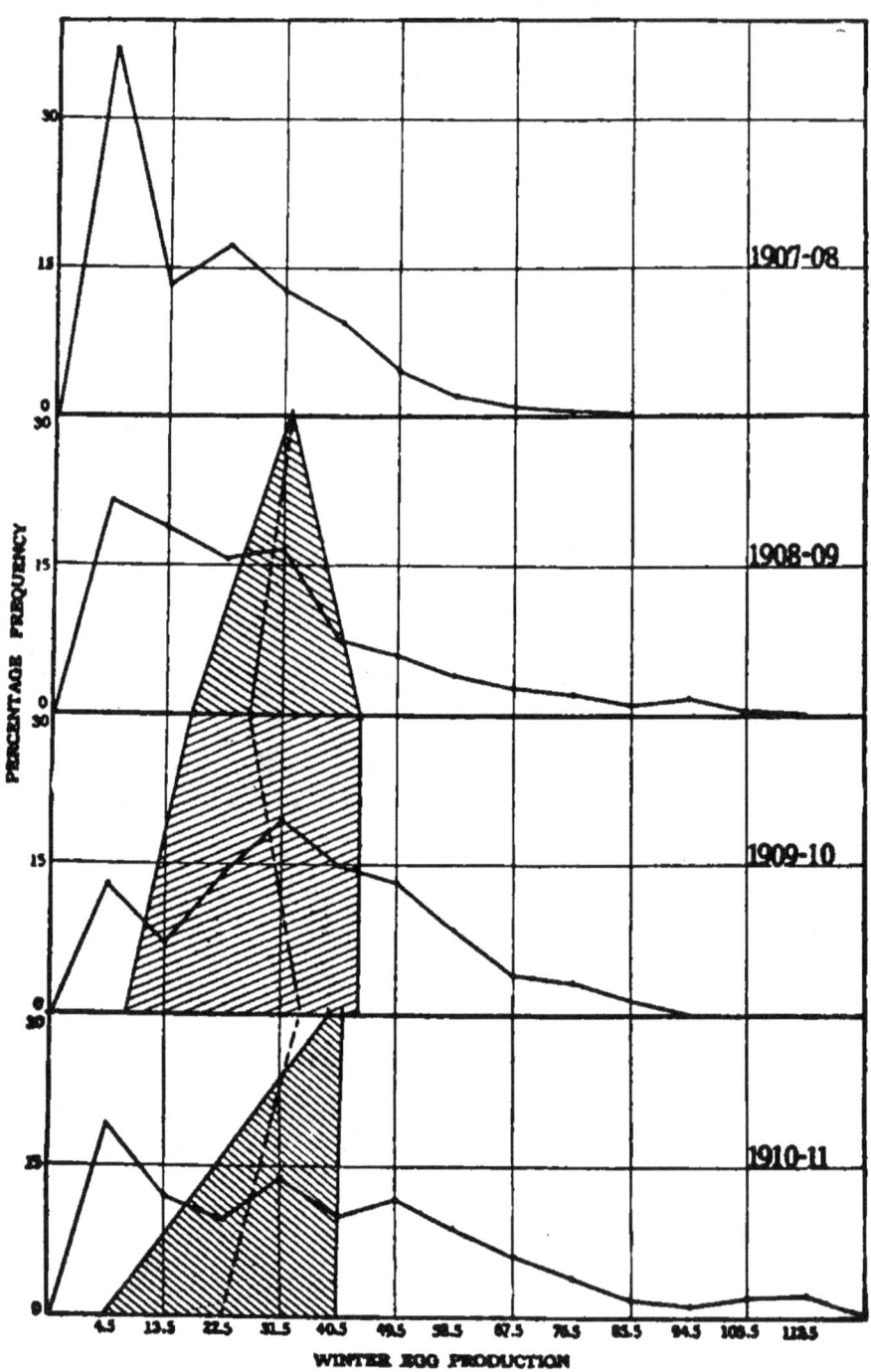

Fig. 82. Diagram showing range of variation and mean fecundity in each generation of line D65D366. Significance of lines and cross hatching as in Fig. 81. q. v.

other led to the production of six adult daughters, all of which are relatively high layers, with the single exception of G495, which has a record of only one egg, and that record is doubtful. This bird has probably never laid an egg, and almost certainly is pathological.

Leaving this bird out of account because pathological, the mean winter production of the family is 52.8 eggs, very strikingly different from the average (9.67 eggs) of the birds of the same generation in the main low line in which the mutation appeared.

Two other daughters of the mutant E248 were mated to ♂ D31, a bird known not only to belong to a genotype of mediocre to low fecundity, but to be remarkably prepotent in respect to this character, so that practically regardless of the females with which he has been mated the get has been uniformly poor in respect to egg production. Four adult females resulted from the two matings under discussion. They have an average winter production of 23.75 eggs. There are several possible explanations of this result, but the most probable is that we have here simply one more instance of the extraordinary prepotency of ♂ D31.

The last of the daughters of the mutant was mated to a cross-bred male, No. 578, and consequently the progeny can not fairly be compared with the pure Barred Rocks in respect to fecundity.

The facts here briefly discussed are shown in the following table and graphically in Fig. 83.

It is apparent from the table and the diagram that the main line and the "mutant" line are entirely distinct. Indeed they do not overlap in their ranges even excepting only the pathological individual G495. The "mutant" pullet E248, for some reason or other, possessed the capacity both to lay a relatively large number of eggs, and the genes necessary to make this quality appear in her progeny. Whether this individual is to be regarded as a true "mutation" would appear to be largely a question of definition. In the writer's opinion the most probable explanation is that E248 is a Mendelian segregation product. That is, let it be supposed that both D168 and D61 were heterozygous with respect to degree of fecundity, and were producing in some (unknown) ratio both "high fecundity" and "low fecundity"

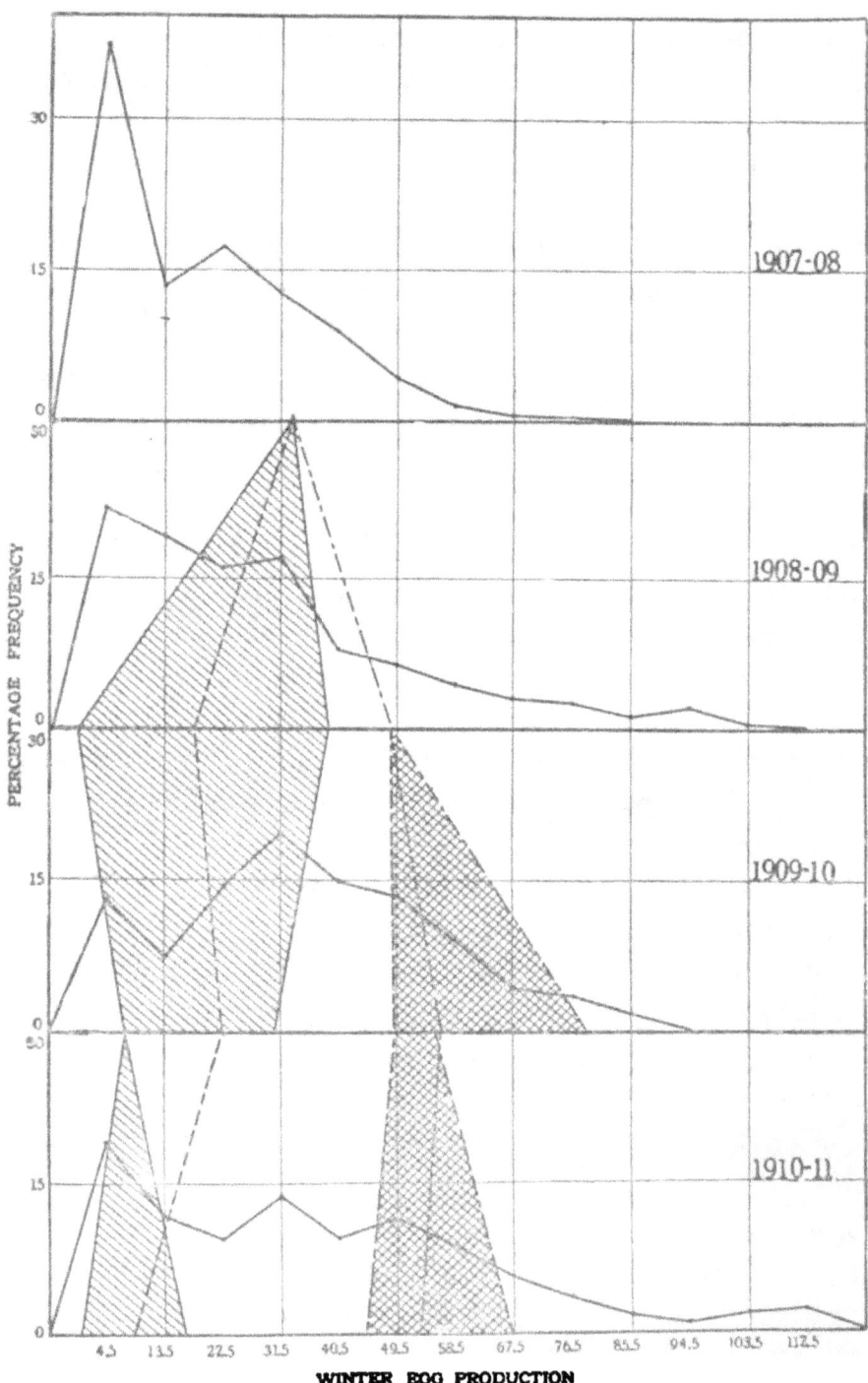

Fig. 83. Diagram of pedigree line D61D168. The significance of lines is the same as in Figs. 81 and 82, except that the mutant line is double cross hatched. For the sake of simplicity E495 and the daughters of D31 are omitted in the 1910-11 generation.

Pedigree Line D61 D168 (Complete)

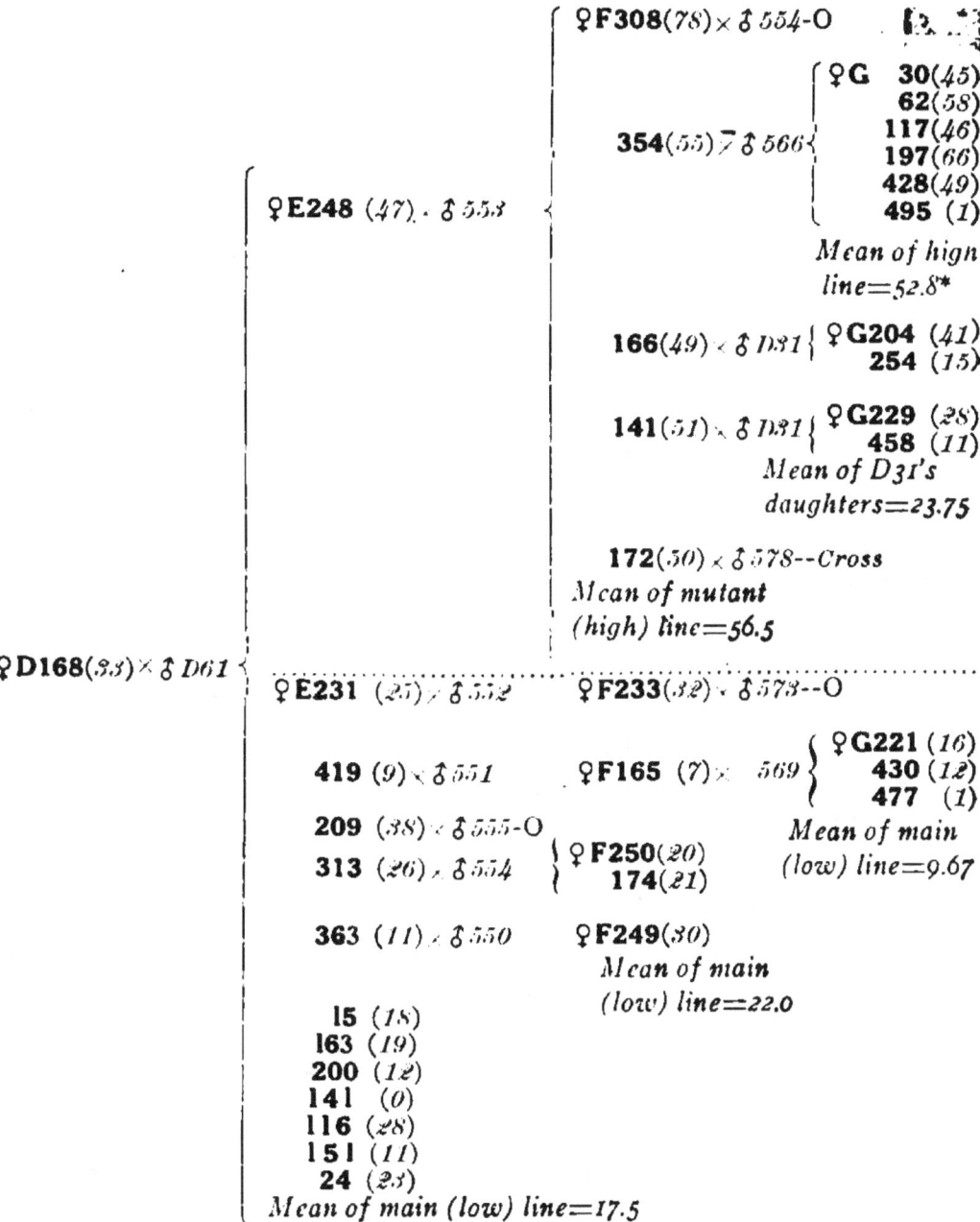

gametes. Then E248 may be supposed to have originated from the union either of two "high fecundity" gametes or one high and one low fecundity gamete. She then would be either a DD or a DR bird, on the assumption, which the facts seem to support, and which I have more fully discussed elsewhere,** that high fecundity is dominant over low.

---

* Omitting G495. See text.

** "Inheritance in 'Blood Lines' in Breeding Animals for Performance, with Special Reference to the '200-egg' Hen," Rept. Amer. Breeders' Assoc., Vol. VI, 1911 (in press).

The subsequent breeding history of E248 indicates that it was probably a DD bird, though the reasons for this opinion can not be fully gone into here. The general view, recently emphasized by Nilsson-Ehle,* that phenomena of mutation are, in many cases at least, merely cases of Mendelian segregation has much evidence in its favor.

The pedigrees which have been given are merely illustrations. Many other similar ones might be cited from the records in hand did space permit. In the experiments during the past three years the attempt has been made to propagate separately lines of high, medium and low fecundity. In the course of this work it has been found that lines of high fecundity were nearly if not quite as likely to have originated with individuals of a low record of production as with those of a high record. Similarly, many low fecundity lines have originated with individuals which were themselves exceedingly high layers. Indeed one of the highest winter layers which have ever appeared in the stock evidently belonged to a genotype of very low fecundity, since it has never been able to produce progeny of anything but the poorest laying capacity. The breeding history of this bird (D352) is indeed so interesting that it may be briefly discussed here. This bird in her pullet year laid 98 eggs between November 10 and March 1 and made a record for the year of over 200 eggs. She was mated and produced plenty of eggs during the hatching season, but they hatched very badly. Only one female worth putting in the house was obtained. This pullet (E356) made a winter record of only 39 eggs, just about the general flock average. E 356 was not mated. Her mother (D352) was kept over and bred to another male the next year, in the hope that as a fowl she might produce more and better chickens than she had as a pullet. As a matter of fact she was again able to produce during the whole breeding season only one pullet worth putting into the laying house. This pullet (F163) made a winter record of but 11 eggs. F163 was bred in 1910, but produced only one daughter worth saving. This daughter, G429, has made a winter record of 18 eggs. It would be hard to get clearer evidence than that afforded by this breeding history that

---

* Nilsson-Ehle, H., "Kreuzungsuntersuchungen an Hafer und Weizen," *Lunds Univ. Arsskr.*, N. F., Afd. 2, Bd. 5, Nr. 2, 1909, pp. 1-122.

D352 belonged to a low fecundity genotype, in spite of her individual high laying record.

THE EFFECT OF THE SELECTION OF FECUNDITY GENOTYPES.

Let us now consider the bearing of the results so far set forth on the problem of selection. Taking first the question of the effect of selection for fecundity within a population it is plain that if different degrees of fecundity have a genotype basis, as the facts above presented and a considerable mass of data of a similar kind, which owing to lack of space can not be given here would appear to indicate, then the results following selection will depend entirely upon the genotypic constitution of the population. If high fecundity genotypes are present they can be isolated by selection. If they are not present selection of high laying hens will not change the average production of the flock.

The aim of the selection experiments since 1907 has been to discover and propagate separately genotypes of high fecundity and genotypes of low fecundity, all the birds being taken from the same general flock. The results of this work are shown in the following table and in Fig. 84. This table is to be regarded as a continuation of that given on p. 156, *supra*, which shows the results of mass selection for high fecundity in the same stock.

Effect of Selection for Fecundity within the Population

1907-08.	Mean winter production of general population..........	15.92
1908-09.	Mean winter production of all high fecundity lines.....	54.16
1908-09.	Mean winter production of all low fecundity lines,.....	22.06
1909-10.	Mean winter production of all high fecundity lines......	47.57
1909-10.	Mean winter production of all low fecundity lines......	25.05
1910-11.	Mean winter production of all high fecundity lines......	50.58
1910-11.	Mean winter production of all low fecundity lines......	17.00

The results indicate the effectiveness of this method of selection. It should be understood, of course, that only those pedigree lines are included in the high line averages which uniformly *in each generation* show high fecundity. A similar consideration applies to the low line averages.

Let us now consider briefly the question of the effectiveness of selection *within* the genotype. According to the "pure line" concept we should not expect selection of high or low individuals belonging to the same genotype to produce any effect, except in cases where segregation has occurred and the selected individuals are really gametically different, though having the same pedigree. An example of this sort has been given in the case of line D61D168 (cf. p. 165, *supra*). The ineffectiveness of selection within the line when something of this sort does not occur is illustrated by line D56D407. In the F_1 generation in this line there were four birds, of which three were good layers and one was a poor layer. Two of the good layers and

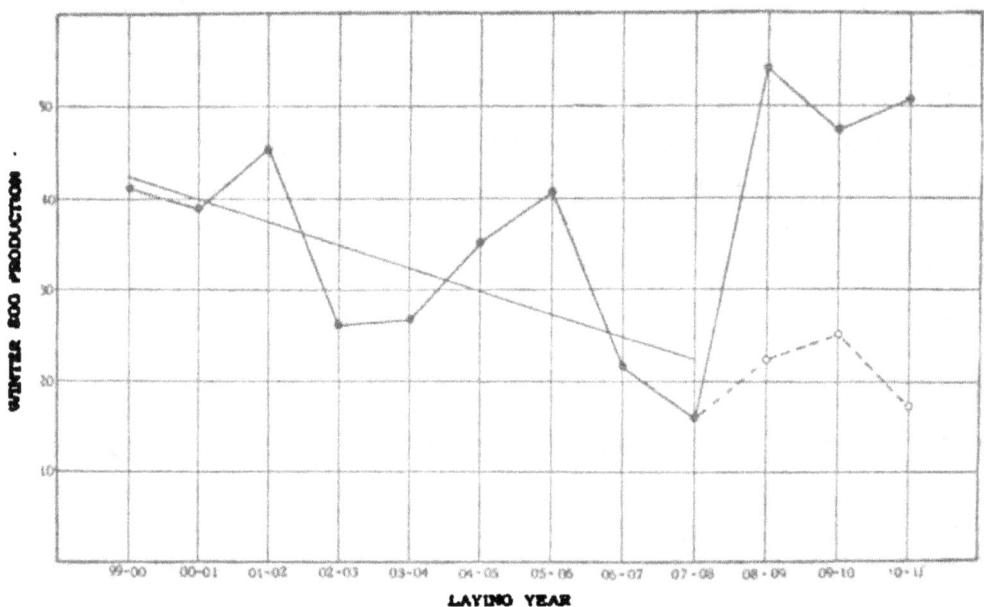

FIG. 84. Showing the effect of selecting high and low fecundity on a genotypic basis. The solid line denotes means of all "high lines;" the dotted line means of all "low lines." Up to 1907-08 the attempt had been to increase egg production by breeding merely from the highest layers, regardless of pedigrees. In 1907 and subsequent years the attempt has been to isolate genotypes of high and low fecundity which shall breed true, each to its own type.

the poor layer were bred. Large families were reared in F_2 and F_3. The average results in the three generations are given in the following table.

Effect of Selection of Good and Poor Winter Layers in the Same Line, D56D407

Generation	F_1	F_2	F_3
Mean winter record of *good* layers and their progeny	76.0	46.7	35.57
Mean winter record of *poor* layers and their progeny	26.0	52.0	36.75*

It is evident that selection within the line here was quite without effect.

Another example of the same thing from line D31D447 may be given by way of further illustration. In this line there was in the F_1 generation a family of ten daughters. Of these some were very good and some were poor layers. All were bred. The mean results are shown in the next table.

Effect of Selection of Good and Poor Winter Layers in the Same Line, D31D447

Generation	F_1	F_2	F_3
Mean winter record of *good* layers and their progeny	62.5	23.75	22.00
Mean winter record of *poor* layers and their progeny	32.0	28.75	14.75

Here again it is plain that selection within the line was without effect. Many more examples of the same sort might be given from the records did space permit. In general there is no evidence whatever that the selection of individuals of different laying records, but belonging to the same fecundity genotype, produces any definite or permanent effect whatever.

DISCUSSION AND CONCLUSIONS.

Taking into consideration all the facts which have come out of this study, one is led to the following view as to the composition of a flock of fowls in respect to fecundity. In the average flock we may presume that there will probably be represented a number of fecundity genotypes, some high, some low, and some intermediate or mediocre. In an ordinary flock these genotypes will be greatly mixed and intermingled. Further, the facts in hand indicate that the range of variation in fecundity *within* the genotype is relatively very large, nearly as great, in fact, as in the general population. Thus while fecundity genotype *means* may be and usually are perfectly distinct, there

---

*If one family of four birds, which ought not in fairness to be included here because they were extremely inbred (brother-sister mating) in connection with another experiment, is excluded this average becomes 49.0.

is much overlapping of individuals in the different lines. In consequence it results that the egg record of an individual bird is of almost no value in helping to tell in advance of the breeding test to what fecundity genotype it belongs. Essentially this same fact has been brought out in all of the work which has been done with pure lines. The only difference in the present case lies in the fact that the range and degree of variation within the line appears to be relatively greater in the case of fecundity than in the case of most characters hitherto studied, as, for example, size relations in beans or *Paramecium*.

The most serious difficulty which confronts one in the attempt to analyze the inheritance of a character like fecundity lies in the almost inextricable mingling of genotypes in the great majority of individuals. This, of course, is a direct consequence of the manner of reproduction. The germ plasm of two separate individuals must unite to form a new individual. By prolonging incestuous mating one may in theory come indefinitely close to reproductive purity, but in practice even this is extremely difficult, if not impossible of accomplishment on any large scale or through any long period of time. The fact simply is that a "pure line" in the strict sense of Johannsen * can not by definition exist in an organism reproducing as the domestic fowl does. This, however, by no means indicates that the inheritance of fecundity does not rest on a genotype basis, or, in other words, that fowls do not carry definite genes for definite degrees of fecundity.

We touch here upon an important point; namely, the relation of the mode of reproduction to the mode of inheritance. As one reflects upon the matter it becomes clear that it is only in the sense of a *reproductive* line that we can not, by definition, have pure lines in organisms where the sexes are separate. It is perfectly possible to have a line of such organisms in which all the individuals are *gametically* pure with reference to any particular character. For example, it is the simplest of matters to establish a line of horses pure in respect to chestnut coat color. Any individual in such a line mated to any other will

---

* Johannsen's definition is as follows: "Mit einer reinen Linie bezeichne ich Individuen, welche von einen einzelnen selbstbefruchtenden Individuum abstammen." ("Ueber Erblichkeit in Populationen und reinen Linien," p. 9.)

never produce anything but chestnut offspring. So similarly with any other character, it is only necessary to obtain homozygous individuals in respect to any character in order to form a gametically pure strain with reference to that character.

It must further be kept clearly in mind that a reproductive "pure line" (in the sense of Johannsen's definition) may be made up of individuals *not* gametically pure (i. e., homozygous). Thus suppose one crosses a yellow and green pea and then takes an F_2 heterozygote individual seed which originated from a self-fertilized F_1 individual as the "single, self-fertilized individual" with which to start a line. The individual which starts such a line arose by self-fertilization and is selfed to produce progeny and would thus fulfil every requirement of a *reproductive* "pure line" as defined by Johannsen. Yet it would produce both yellow and green offspring. On the other hand, as already pointed out, a line which is not, and from the nature of its mode of reproduction never can be, reproductively "pure" may be gametically so (i. e., have none but homozygous individuals with respect to any character).

We then see that the fact that in fowls the sexes are separate and we therefore can not have reproductive "pure lines" gives, *per se*, no reason to suppose that fecundity is not inherited on a genotypic basis. We have to consider the problem of genetic or gametic purity. Do we have homozygote lines in such cases as those discussed in this paper? It plainly is the fact that one can get lines of birds which, broadly speaking, will breed true (perhaps throwing occasionally a few individuals not true to the type of the line) to definite degrees of fecundity. The same thing is true of milk production in dairy cattle, speed in race horses, etc. What are these lines gametically? Theoretically the formation of gametically pure (homozygote) lines with respect to definite degrees of fecundity is simple. Practically it is exceedingly difficult to do this, owing to the fact that (*a*) the character studied is not expressed in the male, and (*b*) it is subject to a wide fluctuating variability caused by environmental conditions. The question as to the gametic constitution of the fecundity lines here discussed obviously can not be answered finally now. It is a matter for much further research. One may, however, form a general conception of the probable gametic constitution of such lines, which has much evidence in its support. The essential points in such a conception are:

1. Probably no line yet obtained is absolutely pure gametically in respect to fecundity. It represents a mixture of a greater or less number of fecundity genes.

2. Lines which breed reasonably true to a definite degree of fecundity may in most cases be taken to be made up of individuals bearing a preponderant number of genes of the particular degree of fecundity to which the line breeds true, so that in gametogenesis a great majority of the gametes formed carry only these genes. They also carry some genes of higher, or lower fecundity, or both kinds. When individuals of a definite (*c. g.*, "high") line thus constituted are bred together the majority of the offspring will, purely as a matter of chance, be produced by the union of two high fecundity gametes. It is quite possible that with families of the size obtained with poultry nearly or quite every individual produced in the line for several successive generations may be of this kind. In the long run, however, it is to be expected that a small number of "off" individuals will appear in the line. These originate by the chance union of two low fecundity genes, or by the union of a "high" gene with a "low" gene of great potency (as in the case of D_{31}, cf. p. 163).

3. The degree to which such a line will breed true will depend upon the proportion of genes of one type (or of very similar types) present. The higher such proportion the less frequently will the "off" individual segregate out. The practical goal to be worked towards is, of course, to obtain several lines not closely related, but all made up only of individuals homozygous with respect to either high or low or any other definite degree of fecundity.

Whether a given degree of fecundity is to be regarded as a single unit character, in the Mendelian sense, or, on the other hand, as a complex dependent upon a particular combination of separately segregable unit characters, can not yet be determined. Every one must recognize the fundamental importance of the investigations of Nilsson-Ehle, Baur and East, which have shown that many characters which at first glance do not appear to conform to any determinate law of inheritance are really complexes, formed by the combination of a number of unit characters, each of which segregates and otherwise behaves in a perfectly regular and lawful manner. There are some facts

PART V.

General Summary.

It is the purpose of this bulletin to present in summarized form the essential results of the experiments in breeding poultry for egg production which have been carried on at the Maine Station during a period of 13 years. These results may be briefly stated here as follows:

1. An experiment in which the highest laying hens were used as breeders showed that mass selection for high egg production on the basis of the trap nest record of the individual alone did not, as a matter of fact, result in a steady continuous improvement in average flock production, even though it was continued for a period of ten years.

2. A further experiment along the same line showed that the daughters of "200-egg" hens with from six to nine years selected ancestry (on the basis of trap nest records) behind them were no better layers, on the average, than birds bred from the general flock.

3. There is no evidence that either (a) the method of housing, or (b) of feeding, or (c) the fact that the chicks were throughout the period of the experiment hatched in incubators and reared in brooders, or (d) the fact that some degree of inbreeding was practiced during the mass selection experiment had anything whatever to do with the outcome of that experiment. It is specifically shown in this bulletin that during the period of selection the adult mortality decreased. It is further shown that at the present time, in spite of the fact that there has been no change in the method of hatching and rearing by artificial means, the records of hatching and of chick mortality are such as to give no indication whatever that the strain of Barred Plymouth Rocks which has been used in all the work in breeding for egg production has become in any way deteriorated through the action of environmental or other factors. It is further specifically shown, by an experiment in out-crossing involving a large number of individuals, that the infusion of new blood into this stock failed to produce any change in the egg production of the progeny. Such a result makes it impossible to suppose that the degree of inbreeding practiced during the mass selection experiment can have had anything whatever to do with the results of that experiment.

which indicate that high fecundity is a character of this kind, but it will require prolonged analysis to decide this, because of the numerous practical difficulties which attend the study of fecundity.

A great help in this analysis, as well as a contributory line of evidence of much weight in supporting the general conception of the manner of inheritance of fecundity set forth above, is derived from the study of crosses between breeds of poultry in which high and low degrees of fecundity are definite breed characters. Studies of this sort carried out at the Maine Station indicate that the relatively high fecundity characteristic of the Barred Rock breed is inherited as a sex-limited character. In this respect it behaves like a simple unit character, but this does not necessarily prove that it is not a complex. More data are needed to settle this point. Of much significance is the fact that, whether simple or complex, fecundity is shown by these experiments in cross breeding to be a character resting on a definite gametic basis.

In conclusion, I think it may fairly be said that the investigations here reported show in the first place that different degrees of fecundity *are* inherited in the domestic fowl, and in the second place, that in all respects wherein it has been possible, considering the inherent difficulties of the material and the character dealt with, to make the test, the method of this inheritance is in entire accord with Johannsen's concept of genotypes.

4. In the laying year 1907-08 a new plan of breeding was adopted as a working hypothesis to be tested by experiment. This plan is based on the employment of individual pedigree records and has its theoretical foundation in the genotype concept of Johannsen. This working hypothesis is fully explained in Part III of the present bulletin. It involves the following factors:

(a) That the egg record of an individual hen gives no definite indication whatever as to what the probable laying of her daughter will be. Examination of hundreds of pedigree records leaves no doubt as to the truth of this fact. Individual birds with high egg records are as likely as not to produce daughters that make poor egg records and vice versa. From the laying record of an individual hen it is quite impossible for anyone to tell whether its progeny will be good layers or poor layers.

(b) A flock of hens, no matter how "pure bred" it may be, is really not a homogenous, unitary aggregation, but instead it is made up of a varying number of lines or strains, each of which tends to breed true to a certain or definite degree of egg productiveness or fecundity. In other words such a flock is a mixture of several component lines. The individuals in each line tend to produce offspring true to the type of the line rather than to the type of the population as a whole, excepting in cases where by chance the population type and the type of one or more lines happen to be the same.

(c) When mass selection alters the population type it does so by a process of isolating from the mixture certain strains whose own types are different from the original general population type and which differ in the direction toward which selection is made. The thing to be sought then in the practical breeding of poultry for increased egg production is to discover by means of pedigree analysis those individuals of the general flock which possess high fecundity in inheritable form. These individuals may then be isolated and propagated and improvement thus brought about.

5. It is shown that by the application of this new plan of breeding it has been possible to isolate from the same stock of birds, which was used in the mass selection experiment, pedi-

gree lines or strains which for four generations (the time covered by the experiment to date) have bred uniformly true to definite degrees of egg production. In this work there have been isolated and are now being propagated lines carrying high egg productiveness, and lines carrying low productiveness, the character apparently being definitely fixed in the pedigree line or strain in each case.

6. In order to determine the mechanism by which fecundity is inherited more data are needed. From the evidence in hand, however, it appears to be the case that this character is inherited fundamentally according to Mendelian principles, though it is not yet clear as to what may be the number and nature of the factors involved. There is, however, clear evidence that high fecundity and low fecundity segregate definitely following crosses between breeds of poultry bearing these characters as definite breed characters. Further studies on this phase of the problem are now in progress.

www.ingramcontent.com/pod-product-compliance
Lightning Source LLC
Chambersburg PA
CBHW062335220526
45469CB00008B/2726